Introduction to Optimization-Based Decision-Making

Series in Operations Research

Series Editors

Malgorzata Sterna, Bo Chen, Michel Gendreau, and Edmund Burke

About the Series

The CRC Press Series in Operations Research encompasses books that contribute to the methodology of Operations Research and applying advanced analytical methods to help make better decisions.

The scope of the series is wide, including innovative applications of Operations Research which describe novel ways to solve real-world problems, with examples drawn from industrial, computing, engineering, and business applications. The series explores the latest developments in Theory and Methodology and presents original research results contributing to the methodology of Operations Research, and to its theoretical foundations.

Featuring a broad range of reference works, textbooks and handbooks, the books in this Series will appeal not only to researchers, practitioners and students in the mathematical community, but also to engineers, physicists, and computer scientists. The inclusion of real examples and applications is highly encouraged in all of our books.

Rational Queueing

Refael Hassin

Introduction to Theory of Optimization in Euclidean Space

Samia Challal

Handbook of The Shapley Value

Encarnación Algaba, Vito Fragnelli and Joaquín Sánchez-Soriano

Advanced Studies in Multi-Criteria Decision Making

Sarah Ben Amor, João Luís de Miranda, Emel Aktas, and Adiel Teixeira de Almeida

Handbook of Military and Defense Operations Research

Natalie Scala, and James P. Howard II

Understanding Analytic Hierarchy Process

Konrad Kulakowski

Introduction to Optimization-Based Decision-Making

João Luís de Miranda

For more information about this series please visit: https://www.routledge.com/ Chapman--HallCRC-Series-in-Operations-Research/book-series/CRCOPSRES

Introduction to Optimization-Based Decision-Making

João Luís de Miranda

CRC Press
Taylor & Francis Group
Boca Raton London New York

CRC Press is an imprint of the
Taylor & Francis Group, an **informa** business

A CHAPMAN & HALL BOOK

First edition published 2022
by CRC Press
6000 Broken Sound Parkway NW, Suite 300, Boca Raton, FL 33487-2742

and by CRC Press
2 Park Square, Milton Park, Abingdon, Oxon, OX14 4RN

Library of Congress Cataloging–in–Publication Data

Title: Introduction to optimization-based decision making / Joao Luis de Miranda.
Description: First edition. | Boca Raton : C&H\CRC Press, 2022. | Series: Chapman & Hall/CRC series in operations research | Includes bibliographical references and index.
Identifiers: LCCN 2021021796 (print) | LCCN 2021021797 (ebook) | ISBN 9781138712164 (hardback) | ISBN 9781032119779 (paperback) | ISBN 9781315200323 (ebook)
Subjects: LCSH: Decision making--Mathematical models. | Operations research. | Mathematical optimization.
Classification: LCC T57.95 .M565 2022 (print) | LCC T57.95 (ebook) | DDC 658.4/03--dc23
LC record available at https://lccn.loc.gov/2021021796
LC ebook record available at https://lccn.loc.gov/2021021797

ISBN: 9781138712164 (hbk)
ISBN: 9781032119779 (pbk)
ISBN: 9781315200323 (ebk)

DOI: 10.1201/9781315200323

Typeset in Minion
by Deanta Global Publishing Services, Chennai, India

To my Family

Contents

Foreword

FEW REALIZE THAT OPTIMIZATION PLAYS A CRITICAL ROLE in decision-making in the complex world in which we live today. And decision-making problems with significant monetary consequences are optimized in order to maximize the reward, or minimize the cost involved; such applications often attract the attention of the highest political circles.

As an example, I cite the problem faced by crude oil producing countries, that of determining how much crude oil they should produce annually; discussed in Chapter 3 of my textbook on the crude oil industry *Models for Optimum Decision Making* published by Springer in 2020.* As explained there, in 1985 the price of crude oil was US $31.75/barrel. Saudi Arabia increased their production by 3% hoping to increase its revenue from crude oil exports, but instead in a few months the price of crude oil bottomed out at US $9.75/barrel. In 2014, the US government used these facts and staged a repeat of an event similar to when Russia was punished for occupying Crimea.

In this book, Professor João Luís de Miranda focuses on describing to young students at various levels, how optimum solutions are obtained for several decision-making applications dealing with optimum utilization of scarce resources using linear programming, integer linear programming, and nonlinear and stochastic models.

I came to know about his effort to prepare this textbook in 2009 when I visited him on a Fulbright senior specialist award at his invitation. He was using the early drafts of this textbook for teaching his courses on decision-making with optimization as its goal in the School of Technology and Management at the Polytechnic Institute of Portalegre, Portugal.

---

* Other interesting publications of Professor Murty can be found at his personal webpage, http://www-personal.umich.edu/~murty/.

He introduces the concepts and algorithmic techniques through illustrative examples in this text. I commend this effort by Professor Miranda in preparing this textbook to teach modelling skills and algorithms to students and send my best wishes for the book's success in the market.

Katta Gopalakrishna Murty
Ann Arbor, Michigan, USA

Preface

DECISION-MAKING HAS BEEN THE OBJECT OF SIGNIFICANT CHANGE in the last few decades, while the Internet has paved the way for a new communication era and mobile technologies have made available information buckets to individuals. Optimization techniques are being revisited, and the mathematical basis of decision-making is playing a critical role in new approaches to artificial intelligence, data science, high performance computing, and other advanced areas.

In the next few years, a rapid expansion in data infrastructures and high-performance computing is expected, and some type of technology federation is foreseeable for both individual users and social networks. Nowadays, more sophisticated educational tools—e.g., augmented reality, virtual reality, big data analytics, and gamification—are being utilized, and distance learning features have been widely promoted due to the COVID pandemic. However, conventional classrooms continue to be part of everyday life in many institutions around the world.

This enhanced book on optimization-based decision-making addresses a diversity of theories and applications, covering basic materials while providing a more focused approach to selected optimization topics. The mathematical level of the text is directed to the post-secondary reader or the university student in the initial years; for that, some mathematical tools are introduced as needed. This lean approach is complemented with a problem-based orientation and a methodology of generalization/reduction. In this way, the book can be useful for students from the science, technology, engineering, and mathematics (STEM) fields, economics and enterprise sciences, social sciences and humanities, and for the general reader interested in multi/transdisciplinary approaches through simple problems.

The text materials include optimization applications that support decision-making, and students from basic cycles up to graduation degrees are targeted through simple problems. The significance of scarce resources, optimality, and marginal values is illustrated by a simple linear algebra (LA) problem that evolves toward an integer linear programming (LP) model. With the LP modeling and calculations simplified, students have the opportunity to recognize more complex problems and are empowered to solve them. Namely, LP duality concepts, Lagrange multipliers through differential calculus, and LP post-optimality analysis are addressed. Additionally, important concepts on integer LP, Game Theory, decision-making under uncertainty, stochastic LP, and robust optimization (RO) are also treated.

Chapter 1 describes several applications of decision-making directed to school education, illustrating them through different optimization instances in a way that is both simple and instructive. The optimal utilization of scarce resources is illustrated though a simple integer LP problem and its instances of limited dimension. The difficulty level for those instances increases from the basic level to the secondary level, as well as solving capacities grow too.

In Chapter 2, important LA tools that are widely used to solve linear systems of equations are presented; while overviewing the types of linear systems, they are also associated with the solutions types, the inverse matrix, and determinants. A graphical approach is also presented, enriching the LA-based decision tooling and complementing this comprehensive insight into linear systems of equations.

Chapter 3 addresses the LP basics; namely, from the graphical approach to the algebraic form, the LP Simplex tableau, to the matrix form. The prior LA approach is enlarged, with the LP Simplex method addressing a linear system of equations at each iteration; and also by obtaining new attributes that improve decision-making and that are directly related to slack variables, dual values, and sensitive parameters.

In Chapter 4, duality and the complementary dual problem are introduced; the transformation from the primal LP problem, the optimal dual solution obtained from the Simplex tableau form, the main properties of dual solutions, as well as the related economic interpretation are discussed too. Note that the LP primal problem is complemented and enlarged to the non-feasible region through the associated dual problem. A new insight concerning optimality analysis and sensitive parameters is thus obtained, and decision-making's attributes are improved once again.

Lagrange multipliers and their relations with the LP dual values are described in Chapter 5, while both calculus and LA procedures are used to obtain the optimal solution for the problem at hand. Firstly, basic notions on differential multivariate calculus are revisited, in particular for the general reader; secondly, a simple optimization problem is presented, namely, the maximization of a two-variable function with one restriction; then, a generalization occurs to consider the maximization of an n-variable function, while m restrictions are simultaneously satisfied. The Lagrange function for the problem is built, and the optimal solution is obtained through Cramer's rule; finally, the Lagrange solution is compared with the LP dual solution.

Chapter 6 addresses the most important procedures for LP optimality analysis, namely the objective function coefficients, the right-hand side (RHS) parameters, and the introduction of new variables. Based on LP Simplex procedures, a sensitivity analysis is developed in a systematic way, and a simple parametric analysis is also performed. The main results are the optimality and feasibility ranges, respectively, for the objective function coefficients and the RHS parameters; the mapping of alterations to the optimal solution with the evolution of pertinent coefficients or parameters is also important. Therefore, the decision maker gains access to new information sources, and these new insights can enhance the response to external fluctuations in prices, resources availability, technology changes, and other uncertainty factors.

In Chapter 7, the important topic of integer linear programming (ILP) is introduced, which provides (i) qualitative improvements due to more realistic solutions, namely when decision variables require integer values; and (ii) better modeling capabilities, for example, with binary variables formulating contingency decisions or fixed charge costs. However, such qualitative improvements come at a cost of quantitative reductions: the LP optimal value is the hard limit for the branch-and-bound (B&B) method, considering also that the solutions space is successively reduced and the objective value is successively cut. In the opposite sense to the generalization approach, the B&B follows a reduction approach that quantitatively constrains the objective function; however, both the modeling features and the integrity attributes largely enhance decision-making through ILP.

A brief insight into Game Theory is provided in Chapter 8, enlarging the decision framework by simultaneously addressing two decision makers, or the two players as they are commonly named. Beyond the key notions of the two players constant-sum game and zero-sum game, the associated LP formulations are developed; the duality approach associated with such LP complementary versions is also addressed, including both the max-min and min-max versions. The LP treatment of both mixed strategies and dominant strategies is an important enhancement, since it directly supports decision-making.

The main topics of decision-making under uncertainty are presented in Chapter 9, including important theoretical and practical issues. Advanced methods are addressed, both for assisting the decision-making process and for decision modeling in uncertain conditions, and a comparison analysis is also developed. The comparison of alternatives using multiple criteria is crucial in the treatment of large and complex projects, for example, the capacity expansion problem. Decision makers are molding real-world decisions through their multiple attitudes, behaviors, beliefs, and values, and the appreciation of such decision processes requires advanced methods.

Chapter 10 continues the treatment of uncertainty, focusing on robust optimization (RO). The decision maker is assuming probabilistic scenarios, which requires stochastic treatment beyond the typical expectation value. Stochastic programming (SP) and related attributes for decision-making are presented. Noting that the generalization of the deterministic models to the stochastic framework drives either qualitative or quantitative improvements, namely, by promoting robustness both on models and on solutions. A case study is described, including pertinent RO procedures associated with typical economic estimators and industry-based parameters. In addition, tables summarizing the main subjects described in the book are presented, including both the typical methods described in Chapters 1–5 and the more advanced approaches in Chapters 6–10.

With the best motivation to present optimization methods that typically support decision makers, obviously other important applications cannot be included in this book. Namely, the LP special cases related to transportation problems, with networks optimization, or even graph-based problems that can be addressed as specific instances of the minimum cost flow problem. Indeed, the importance of such topics would deserve a dedicated volume, as well as other effective tools for multi-criteria decision-aiding/making.

The book follows a lean approach that integrates three main subjects, namely, introducing optimization problems, noting important decision-making items, and complementing with a generalization/reduction insight. Within this approach, the solution to examples and problems instances are presented without software brands or applications; in fact, a number of solving applications are available on the Internet, and cloud computing is common nowadays.

The materials in this book have been classroom tested over many years, including several periods of teaching mobility when invaluable feedback was received, either from undergraduate or graduate students. Good colleagues from different institutions have provided useful suggestions, and I gratefully acknowledge those very helpful comments, particularly during mobility periods and international exchanges. I also am very grateful to CRC Press and Deanta Global, to the editorial and production staff, for their enduring support, patience, and skill.

Author

João Luís de Miranda is Professor at ESTG-Escola Superior de Tecnologia e Gestão (IPPortalegre) and researcher in optimization methods and process systems engineering (PSE) at CERENA-Centro de Recursos Naturais e Ambiente (IST/ULisboa). He has been teaching for more than 20 years in the field of mathematics (e.g., calculus, operations research [OR], management science [MS], numerical methods, quantitative methods, and statistics) and has authored/edited several publications on optimization, PSE, and education subjects in engineering and OR/MS contexts. João Luís de Miranda addresses the research subjects through international cooperation in multidisciplinary frameworks, and serves on several boards/committees at national and European level.

First Notes on Optimization for Decision Support

THIS CHAPTER DESCRIBES SEVERAL applications of decision-making directed toward school education, toward children's and young students' learning, illustrated through different optimization instances in a simple and instructive way.

1.1 INTRODUCTION

We live in a world with increased pressures on resources, from extraction to distribution, including residues valorization and materials reuse. Thus, the focus is on the optimal utilization of scarce resources, and is illustrated by a simple integer linear programming (LP) problem and related instances. The problem's difficulty grows from basic levels to the secondary level, in tandem with the problem-solving capacities of children and young students.

DOI: 10.1201/9781315200323-1

The problem instances are aimed at different levels of school education, with instances numerically bounded and calculations simplified. There are three basic levels of school education:

- **First level:** From 6 to 9 years old, up to 4 years of school.

- **Second level:** Between 10 and 11 years old, up to 6 years of school.

- **Third level:** From 12 to 14 years old, up to 9 years of school.

A more difficult level is the fourth level (secondary, senior high school; usually 15–17 years old, up to 12 years of school), because this level also relies on computing skills. With greater instances and less solution time, an LP formulation is required, which is developed and a procedure is applied. It must be noted that some places include LP modeling in the mathematics curricula at secondary level. Therefore, the fourth instance provides students with a good opportunity: on the one hand, they are required to treat a more complex problem or instance; on the other hand, they are empowered to obtain the optimal solution for this type of problem, i.e., LP instances.

In addition, the significance of both scarce resources and related marginal values is addressed:

- It consists of either an open-ended scenario or the typical "*What if?*" analysis, where students with different knowledge levels (first, second, or higher levels) are challenged.

- Partial or complementary questions are properly ordered, allowing students to develop their own solution methods, which they perform autonomously.

- Using either teamwork (two or three students in each group) or individually, tutorial supervision promotes the goals at each level, ensuring that they are accomplished.

- The tutorial goals include self-contained strategies that young students can build on to solve the problem instances; for that, the instances are not defined or partitioned in a tight format.

Thus, the topics within the four instances can overlap either the first and second levels or the second and third levels; in addition, it is also possible that a first-level student may attain the final results of the higher levels, and even solve all the problem instances.

1.2 FIRST STEPS

The problem is tailored to the first level and the calculations are simplified using a bounded instance: only integer non-negative numbers are included, and the sums and subtractions do not reach 20.

These first steps require only basic notions to support decision-making, such as enumeration, basic algebraic operations (addition, subtraction, multiplication, division), simple combinatorial counting, and direct comparison of several alternative solutions. Thus, the skills required are elementary, related to the numeric natural system, the four basic operations, and their properties to simplify calculations.

1.2.1 The Furniture Factory Problem: First Level

Noddy is trying to help Big Ears build some tables (each table is worth four chocolate cakes) and some chairs (each chair is worth three chocolate cakes) according to Figure 1.1a.

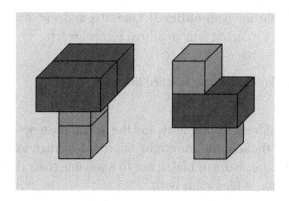

FIGURE 1.1a   Furniture factory problem: First level.

However, there are only eight small-red and three big-blue pieces (Figure 1.1b).

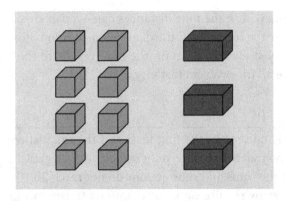

FIGURE 1.1b   Availability of pieces: Eight small-red and three big-blue pieces.

How many tables and chairs does Noddy need to build to earn the most chocolate cakes?

Instance: Eight small-red pieces and three big-blue pieces.

- **Optimal solution:** Zero tables and three chairs; the value is 9 (0 × 4 + 3 × 3) and two small-red pieces are left.

- **Second solution (sub-optimal):** One table and one chair; the value is 7 (1 × 4 + 1 × 3), but four small-red pieces are left.

How do you know you have reached the end?

This question addresses optimality, and the notion of an optimal solution; it occurs when the solution cannot be improved further, as it is not possible to build more chairs or tables, nor to undo one chair (lost value is 3) to make one table (added value is 4).

A set of "*What if?*" questions address the marginal value notion, specifically for the big-blue pieces. Namely,

What if ...

- *There is one more big-blue piece?*

  It could build one more chair (gaining 3), obtaining a new maximum value of 12 (corresponding to 4 × 3).

- *There is again one more big-blue piece?*

  It could build one more table (gaining 4) while undoing a chair (losing 3); therefore, the marginal value is 1 (4 − 3) and the total value is 13, corresponding to one table and three chairs (1 × 4 + 3 × 3).

- *There is another big-blue piece to add?*

  Again, it could build one more table (gaining 4) while undoing a chair (losing 3); therefore, the marginal value remains 1 and the total value is 14 (2 × 4 + 2 × 3); this situation is repeated two more times, until the components of the two other chairs are made available to build two additional tables, obtaining a total of 16 (4 × 4 + 0 × 3) with a marginal value of 1 for each new big-blue piece.

- *There is yet another big-blue piece to add?*

  No more additional tables can be produced, as there are no more chairs to undo; therefore, the marginal value is zero and the optimal solution is the previous solution.

1.3 INTRODUCING PROPORTIONALITY

Proportionality and related notions are preferentially introduced in the second level, while instances data are proportionally increased, and operations with fractional numbers are also introduced. Additionally, comparative reasoning can be stimulated, while the attributes of the optimal solution are considered. In this way, the main goals for first- and second-level education can be achieved.

By introducing simple variations on the instance data and promoting a sensitivity or "*What if?*" analysis, the problem approach is open-ended. These topics can be integrated as complementary questions, asked after the students have correctly reasoned the optimal solution and the marginal values.

1.3.1 The Furniture Factory Problem: Second Level

Noddy is trying to help Big Ears build some tables (each table is worth four chocolate cakes) and some chairs (each chair is worth three chocolate cakes), but Big Ears only has eight small-red and four big-blue pieces (Figure 1.2a and b).

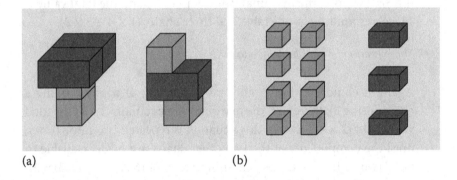

(a)   (b)

FIGURE 1.2   (a) Furniture factory problem: Second level. (b) Availability of pieces: Eight small-red and four big-blue pieces.

If Noddy has 8000 small-red and 4000 big-blue pieces available, how many tables and chairs does Noddy need to build to earn the most chocolate cakes?

Instance: Eight small-red pieces and four big-blue pieces.
The initial phase is similar to the first level's instance, with a reminder to accurately confirm the attributes of both the optimality and the marginal value.

- **Optimal solution:** Zero tables and four chairs; the value is 12 (0 × 4 + 4 × 3) and all the pieces are utilized, both the small-red and the big-blue pieces.

- **Second solution (sub-optimal):** One table and two chairs; the value is 10 (1 × 4 + 2 × 3), but now two small-red pieces are not utilized.

How do you know you have reached the end?

This question is again addressing the optimal solution that occurs when the solution cannot be improved further; namely, when it is not possible either to build more chairs/tables because all the pieces are utilized or to undo one chair (loss is 3) to make one table (gain is 4).

Once again, a set of *"What if?"* questions address the marginal value for the big-blue pieces.

What if ...

- *There is one more big-blue piece?*

  One more table can be built (+4) while undoing a chair (–3); therefore, the marginal value is 1 (4 – 3) and the total value is 13, corresponding to one table and three chairs (1 × 4 + 3 × 3).

- *There is again one more big-blue piece?*

  In a similar way, one more table (+4) can be built while undoing a chair (–3); therefore, the marginal value remains 1 and the total value is 14, as now the optimal solution considers two tables and two chairs (2 × 4 + 2 × 3).

- *There is another big-blue piece to add?*

  This situation can be repeated two more times, until the components from the two other chairs are made available to build two additional tables; then, a total of 16 (4 × 4 + 0 × 3) is obtained, with a marginal value of 1 for each new big-blue piece.

- *There is yet another big-blue piece to add?*

No more tables can be produced, as there are no more chairs to undo; therefore, the marginal value is zero and the optimal solution remains the previous solution, with four tables and a gain of 16 ($4 \times 4 + 0 \times 3$).

The following questions address proportionality and non-proportionality topics.

What if 8000 small-red and 4000 big-blue pieces are available?

Addressing the proportionality topic within the set of 8000 small-red and 4000 big-blue pieces available, the proposed solution is proportional to the previous solution in the initial phase where a set with eight small-red and four big-blue pieces are treated.

- Now, if a scale 1000 times higher is considered, the optimal solution proportionally presents 0 tables and 4000 chairs, totaling 12000 ($0 \times 4 + 4000 \times 3$).

- Note that all the pieces, that is, both the small-red and the big-blue pieces are fully utilized.

And what if 8008 small-red and 4005 big-blue pieces are available?

Note the previous solution, concerning the set with 8000 small-red pieces and 4000 big-blue pieces, can be combined with a first level approach, with the remaining eight small-red pieces and five big-blue pieces.

- The final solution is thus to combine the partial set, 0 tables and 4000 chairs, in the partial amount of 12000, with the partial set, one table and three chairs, in the partial amount of 13, thus gaining 12013 ($1 \times 4 + 4003 \times 3$).

This complementary question addresses non-proportionality and can be considered as an extension issue; however, it promotes additional algebraic procedures, mathematical formulations, and graphical approaches, as presented in the next section.

1.4 A NON-PROPORTIONAL INSTANCE

For the third level, the initial instance is enlarged in a proportional way, and it is assumed that the attributes associated with the optimal solution and marginal values are already well understood. The goal levels consider simple counting principles, algebraic transformations, or function manipulation. Thereafter, a non-proportional instance is introduced.

The problem is open-ended in the sense that data variability can be introduced; this variability transforms the proportional instance into a non-proportional instance. Even in the latter, an approximation procedure can lead to the optimum solution, and young students should be aware of this fact. Another possible extension is introducing the LP formulation and the corresponding graphic representation, which is treated at the secondary level.

1.4.1 The Furniture Factory Problem: Third Level

A furniture factory builds tables at a profit of 4 euros per table, and chairs at a profit of 3 euros per chair. If only eight small-red and five big-blue pieces are available, what combination of tables and chairs needs to be built to make the most profit? (Figure 1.3a and b)

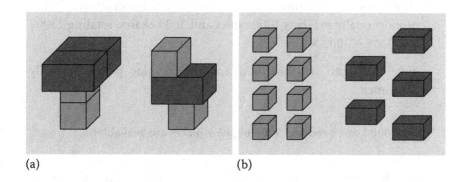

(a)   (b)

FIGURE 1.3   (a) Furniture factory problem: Third level. (b) Availability of pieces: Eight small-red and five big-blue pieces.

> If the availability of the small-red pieces is 8008 and the availability of the big-blue pieces is 5006, how many tables and chairs need to be built to make the most profit?

Instance: Eight small-red pieces and five big-blue pieces.
The initial phase is similar to the second level, as described in the previous section; once again, it is proposed to accurately confirm either the notions of optimality or marginal value, as well as the notion of proportionality. As in the previous sections:

- **Optimal solution:** One table and three chairs; the value is 13 (1 × 4 + 3 × 3) and all the pieces are utilized, either the small-red or the big-blue pieces.

- **Second solution (sub-optimal):** Two tables and one chair; the value is 11 (2 × 4 + 1 × 3), but in this instance two small-red pieces are not utilized.

What if 8000 small-red and 5000 big-blue pieces are available?

Addressing again the proportionality topic within the set of 8000 small-red and 5000 big-blue pieces, the proposed solution is proportional to the previous one in the initial phase:

- Now, if a scale 1000 times higher is considered, the optimal solution proportionally presents 1000 tables and 3000 chairs, totaling 13000 (1000 × 4 + 3000 × 3).

- Note that both the small-red and the big-blue pieces are fully consumed.

What if 8008 small-red and 5006 big-blue pieces are available?

Note that this can be combined with the previous solution, concerning the set with 8000 small-red pieces and 5000 big-blue pieces, with an approach similar to the initial phase, with the remaining eight small-red pieces and six big-blue pieces.

- The final solution thus combines the partial set, 1000 tables and 3000 chairs, in the partial amount of 13000, with the partial set, two tables and two chairs, in the partial amount of 14, now gaining 13014 $(1002 \times 4 + 3002 \times 3)$.

In addition, it is important to formulate the linear equations:

- For the total value, i.e., the contribution of the amounts for tables (t) and for chairs (c):

$$PROFIT = 4t + 3c$$

- as well as the constraints on the usage of big-blue pieces (each table uses two pieces, while each chair only uses one piece, within the total usage of 5006 big-blue pieces):

$$2t + 1c = 5006$$

- and small-red pieces (each table uses two pieces, while each chair also uses two pieces, with a total usage of 8008 small-red pieces):

$$2t + 2c = 8008$$

The point here is to numerically verify the solution obtained by the proportional estimate, by assigning values to variables t and c, and then to compare the two approaches.

A suitable extension follows, by successively supplying another big-blue piece and studying the evolution of

- the optimal solution—with one additional big-blue piece, one table will be built (gaining 4) while undoing one chair (losing 3);
- the marginal value, that is, the added value for one additional big-blue piece is 1 $(4 - 3 = 1)$; and

- to verify these calculations and related solutions, both empirically (using the approximation with big-blue and small-red pieces) and computationally (using calculators or worksheet, and presenting tables or graphs).

Note that exact and complete solutions are assumed from the equations, that is, that there are no pieces left (either big-blue or small-red pieces) when assuming equality relations; therefore, the numerical extension is limited due to the occurrence of non-feasible instances.

1.5 AN ENLARGED AND NON-PROPORTIONAL INSTANCE

At the secondary level, a large and non-proportional instance is presented, so its dimension and resolution time are unlimited. Thus, the students must turn to LP modeling and perform their calculations with equipment support. For this level, the LP framework is typically included in the mathematic curricula, and the students are thus empowered with a tool to solve LP problems.

The supervisor's role is to support the utilization of free tutorial software, usually made available over the internet, which provides the optimal solution for the instance at hand, and also its graphical representation. The main goals include LP modeling; the LP graphic with linear functions represented by straight lines and their interceptions; the study of derivative concepts; and the potential development of an approximation (*greedy*) procedure. Potential extensions include some topics usually performed at graduate level, namely, the notions and reasoning associated with the well-known LP Simplex method; the economic interpretation of dual values; and developing a brief sensitivity analysis.

1.5.1 The Furniture Factory Problem: Fourth Level

A furniture factory builds tables at a profit of 4 euros per table, and chairs at a profit of 3 euros per chair. If only eight small-red pieces and six big-blue pieces are available, what combination of tables and chairs needs to be built to make the most profit? (Figure 1.4a and b)

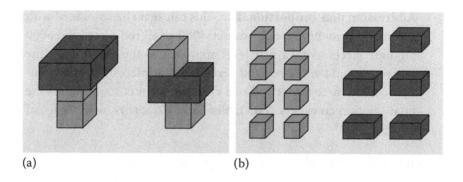

(a)   (b)

FIGURE 1.4   (a) Furniture factory problem: Fourth level. (b) Availability of pieces: Eight small-red and six big-blue pieces.

> If the availability of the small-red pieces is 8008 and the availability of the big-blue pieces is 6007, how many tables and chairs need to be built to make the most profit?

Instance: Eight small-red pieces and six big-blue pieces.
Initial phase: Similar to the third level, it is necessary to accurately confirm the attributes of the optimality and the marginal value, the notion of proportionality, as well as the formulation treatment.

Proposed solution (the same question from the third level): Addressing non-proportionality, it can be combined with the previous solution (8000 small-red pieces and 6000 big-blue pieces) with an approach similar to the first level's instance (with the remaining eight small-red pieces and seven big-blue pieces); the final solution is (2000 + 3) tables and (2000 + 1) chairs, gaining 14015.

- **Optimal solution:** As in the initial steps, two table and two chairs; the value is 14 (2 × 4 + 2 × 3) and all the big-blue and the small-red pieces are utilized.

- **Second solution (sub-optimal):** One table and three chairs; the value is 13 (1 × 4 + 3 × 3), but one big-blue piece is not utilized.

- **Addressing proportionality:** If a scale 1000 times higher is considered, then the optimal solution proportionally presents 2000 tables and 2000 chairs, with a total amount of 14000 (2000 × 4 + 2000 × 3); also note that all the pieces are fully consumed.

- **Addressing non-proportionality:** This can again be combined with the previous solution for the subset, 8000 small-red pieces and 6000 big-blue pieces, with a first level approach to the subset with the remaining eight small-red and seven big-blue pieces; the later solution counts three tables and one chair, the gain is 15, and thus the final solution considers 2003 tables and 2001 chairs, with a gain of 14015.

It is important to formulate the linear expressions:

- for a total profit, such as the contribution of the values of tables (t) and chairs (c):

$$\text{PROFIT} = 4t + 3c$$

- as well as the constraint on using the big-blue pieces (each table uses two pieces, while each chair only uses one, with a total usage of up to 6007 big-blue pieces):

$$2t + 1c \leq 6007$$

- and the constraint on using the small-red pieces (each table uses two pieces, while each chair also uses two pieces, with a total usage of up to 8008 small-red pieces):

$$2t + 2c \leq 8008$$

It is interesting to computationally check the solution resulting from the LP model and the solutions obtained in prior estimates (proportional and counting) by assigning values to the variables t and c and comparing the two approaches. When computational support is available, a suitable extension follows, for example:

- Appreciation of the marginal value of the big-blue pieces; namely, altering the first inequality's right-hand side (RHS) from 6003 to 6009 and keeping constant the remaining parameters, the marginal value is constant and equal to 1.

- Alteration in the number of small-red pieces in the second inequality's RHS; integer solutions are obtained only when two small-red pieces are added, by making it possible to "undo" one table in order to collect all those pieces and produce two chairs; so, it is not convenient to designate such an alteration as a continuous marginal value.

1.6 CONCLUDING REMARKS

Different instances of a well-known problem are proposed for young students at different school education levels, in a manner to challenge them toward alternative strategies and solution methods. Several appealing activities and simple calculations are proposed, which are linked with the typical goals for the respective study level, with the difficulty of instances increasing successively.

The decision-making is illustrated by a simple LP problem, where the main topics include the attributes of the optimality and the marginal value of big-blue pieces. By generalizing the first instance, the proportionality properties of LP are featured in the second instance; additivity properties are also featured in the third instance, either on the objective function or in the linear restrictions; and the divisibility of LP variables is also addressed, as well as graphic approaches and computational empowerment, in the fourth instance.

A generalization approach is observed on numbers and operations. Namely, adding natural numbers, the result is also a natural number, also known as a positive integer; however, when subtracting two natural numbers, the result is a negative integer when the second number is greater than the first number. A similar enlargement can be observed for the multiplication and division of integers: multiplying two integers, the result is also an integer; however, when dividing two integers, the result is a fractional number simply if the second number is greater than the first number. Note also that the multiplication of two equal numbers, for example, and reversing the operation through the square root can lead to a new enlargement onto the real numbers set.

In addition, note that:

- Children and young students in the initial levels (first and second cycle) are also able to obtain the optimal solution for the most difficult instances; however, it is important to alert them to the existing constraints, and to remind them about accuracy, proportionality, and optimality ("no pieces left").

- Students in the most advanced levels (third cycle and secondary) can visualize the problem initially with plastic pieces, and then proceed to mathematical expressions, graphical analysis, and computational treatment, and also compare the solutions obtained through the different approaches.

- In small groups (two or three students), plastic pieces can be useful, either for manipulating or improving visualization; the tutor should only intervene when the progress is stalling , for example, by asking about attributes that may induce the appropriate procedure; the solution time will obviously vary for each student, as well as the depth of the proposed analysis.

The tutorial approach can include different topics, for the reader interested in teaching and learning issues, namely,

- The LP objective (that is, the best possible result), the constraints on the number of available pieces (small-red pieces: 8; big-blue pieces: from 3 to 6), and the related information (the composition of tables and chairs) must be well detailed.

- It must be required that both the final results and the interim calculations are accurate, including the exact numbers of tables and chairs, as well as the total value for the objective function.

- In the furniture factory problem, one single decision maker is assumed, and one alternative is selected using one and only one attribute; additionally, other decision types can be introduced, for example, creating two sets, one with alternatives to select (e.g., solutions above a reference value) and the other with alternatives to reject (e.g., solutions below the same reference value).

- The value (e.g., profit, gain) and the utility notions are implicit in the problem statement, but they can be complemented with other important concepts, such as improvement or increment; additionally, where there is no improvement between two single-attribute alternatives, both the indifference notion and multi-attribute decision-making can be discussed.

- The creation of different sets and alternative distributions can come at the cost of complexity; however, the qualification, appreciation, and evaluation of multi-attribute alternatives can enlarge the discussion.

CHAPTER 2

Linear Algebra

T HIS CHAPTER PRESENTS IMPORTANT linear algebra (LA) tools that are widely used in solving linear systems of equations, overviewing the types of linear systems that may arise, and associating them with the types of solutions, the inverse matrix, and determinants. A graphical approach is presented, enriching the LA-based decision tooling and complementing this comprehensive insight into linear systems of equations.

2.1 INTRODUCTION

Linear algebra is the study of vector spaces and the linear transformation of such spaces. Many problems can be organized and represented by linear systems of equations, which can, in turn, be represented as sets of vectors or matrices. Methods for the linear transformation of vector spaces can be used in the resolution of such problems.

LA can be used to aid decision-making and solve optimization problems. In the following problem, a furniture store makes profit of 40 euros per table (t) and 30 euros per chair (c). Each table is made from two big-blue pieces and two small-red pieces, while each chair is composed of one big-blue piece and two small-red pieces (Figure 2.1).

The problem is to find the optimal number of tables and chairs that can be manufactured from a limited number of small-red and big-blue pieces in order to maximize profit; for example, as shown in Figure 2.2, eight small-red pieces and five big-blue pieces are available in the factory's stock.

The solution can be determined by trial-and-error procedures, as the number of pieces is known, only two types of pieces are addressed

DOI: 10.1201/9781315200323-2

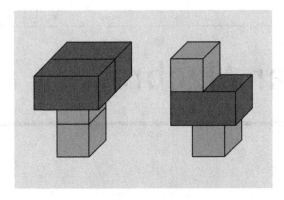

FIGURE 2.1   Furniture factory problem: Composition of tables and chairs.

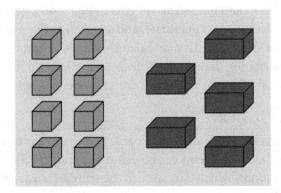

FIGURE 2.2   Availability of pieces; in this instance, eight small-red pieces and five big-blue pieces.

(the big-blue and the small-red), and the number of pieces of each type is manageable. However, a more elaborate approach is required for

- Different combinations of pieces, for example, based on several colors (black and white, gray and orange, green and brown) or materials (metal, plastic, wood).

- A larger number of pieces that make the construction mode unattainable, similar to a counting method that becomes ineffective for large numbers.

- The daily alterations in the number of pieces available in stock that would occur in a real-world application.

Therefore, the trial-and-error approach is not suitable in the production planning of a real-world company that uses many components and parts, incurring wide variations in the available resources, or with a large number of batches. This situation occurs very frequently in the manufacturing industry, for example, in automotive, aerospace, electronics, mechanical equipment, and tools.

The first step is to represent the problem in mathematical form, organizing the information in such a way that a linear system of equations can express the mathematical relationship between all the elements of the problem. In this example, the most significant relationships can be expressed as follows:

$$\text{Maximize \{Total Profit\}} \tag{2.1a}$$

Subject to

$$\begin{Bmatrix} \text{Big-blue pieces} \\ \text{used for} \\ \text{producing tables} \end{Bmatrix} + \begin{Bmatrix} \text{Big-blue pieces} \\ \text{used for} \\ \text{producing chairs} \end{Bmatrix} \leq \begin{Bmatrix} \text{Availability} \\ \text{of big-blue pieces} \end{Bmatrix} \tag{2.1b}$$

$$\begin{Bmatrix} \text{Small-red pieces} \\ \text{used for} \\ \text{producing tables} \end{Bmatrix} + \begin{Bmatrix} \text{Small-red pieces} \\ \text{used for} \\ \text{producing chairs} \end{Bmatrix} \leq \begin{Bmatrix} \text{Availability} \\ \text{of small-red pieces} \end{Bmatrix} \tag{2.1c}$$

To write this problem as a linear model, it is necessary to define:

- **Decision variables:** The decision variables are the elements of the problem that can be controlled, the value of which depends on the decision maker's choice. In this problem, the decision maker selects the quantities of chairs and tables to produce in a suitable time horizon. As such, the decision variables are the number of chairs, c, and the number of tables, t.

- **Objective function:** The objective function calculates a value to indicate how good a solution is. An objective function can be a minimization function or a maximization function. In the example, the objective is to find a solution that achieves the greatest profit possible. As such, the objective function will be to maximize profit, computed as the number of chairs (c) times the profit per chair (30 euros), plus the number of tables (t) times the profit per table (40 euros).

- **Constraints:** The constraints of a problem indicate which solutions are and which are not valid. In this example, the objective is to maximize profit; if the constraints on the decision variables are missing, then the profit is infinite because the number of tables and chairs can also be unlimited. However, an infinite number of furniture items cannot be made because the quantities available for each component are limited. As such, the total amount of small-red pieces used cannot surpass the associated stock availability, and similarly the total amount of large pieces cannot surpass their availability.

Note that in order to be able to solve any problem using LA methods, both the objective function and all of the constraints must be linear expressions. In addition, the availability of pieces and the profit obtained from each product are external elements, their values are out of the decision maker's control, and are known as the problem parameters.

Suppose that the availability of small-red pieces is 6007 and big-blue pieces is 8008. In this instance, the model follows:

$$\max Z = 40t + 30c \tag{2.2a}$$

Subject to

$$2t + 1c \leq 6007 \tag{2.2b}$$

$$2t + 2c \leq 8008 \tag{2.2c}$$

This optimization problem, however, cannot be solved using LA methods as expressed, because each relation represents an inequality instead of an equality. For this reason, before solving this problem using LA methods, the assumption of equality makes sense.

- In fact, it makes sense to use up all the resources available: any solution in which fewer chairs or tables are built would worsen the objective function value, it would be a suboptimal solution; and

- it is not possible to build more chairs and tables than the optimal solution, because no more resources are available.

So, it is clear that the problem solution will be the same if the constraints are expressed as equalities instead of *less than or equal to* inequalities. Then, the problem is satisfactorily represented through linear equations as follows:

$$\max Z = 40t + 30c \qquad (2.3a)$$

Subject to

$$2t + 1c = 6007 \qquad (2.3b)$$

$$2t + 2c = 8008 \qquad (2.3c)$$

Note that this problem presents the same number of equations and variables. This means that only one solution, defining a specific value for each decision variable, exists. This will not always be the case, since a system of linear equations can be

- *underdetermined*, if the system consists of more variables than equations, and so usually the number of possible solutions to the problem is infinite; or

- *overdetermined*, when there are more equations than variables in the system and usually no solution can be found that fulfills all the equations; in such a case, the system is said to be incompatible.

2.2 GAUSS ELIMINATION ON THE LINEAR SYSTEM

Gauss elimination is a method used to solve linear systems of equations, which consists of making linear transformations on the equations system until it takes a form that allows the variables' values to be read directly.

It is important to note that only linear transformations are allowed: linear equations can be combined by adding to and subtracting from others, and they can only be multiplied or divided by scalars. In this case, the transformation starts by subtracting the first equation from the second equation:

$$\begin{cases} 2t + 1c = 6007 \\ 2t + 2c = 8008 \end{cases} \leftrightarrow \begin{cases} 2t + 1c = 6007 \\ (2-2)t + (2-1)c = (8008-6007) \end{cases} \leftrightarrow \begin{cases} 2t + 1c = 6007 \\ 0t + 1c = 2001 \end{cases} \qquad (2.4)$$

The result is a value for the number of chairs, c, which can be substituted in the first equation to obtain the number of tables, t.

$$\begin{cases} 2t + (1-1)c = (6007 - 2001) \\ 0t + 1c = 2001 \end{cases} \leftrightarrow \begin{cases} 2t + 0c = 4006 \\ 0t + 1c = 2001 \end{cases} \leftrightarrow \begin{cases} 1t + 0c = 2003 \\ 0t + 1c = 2001 \end{cases} \quad (2.5)$$

QUESTION 1:

For the linear system:

$$\begin{cases} 2t + 1c = 6007 \\ 3t + 2c = 8008 \end{cases} \quad (2.6)$$

Using Gauss elimination, the correct procedure is

(a) To multiply the first line by (3/2) and subtract it from the second line.

(b) To divide the first line by 2 and multiply it by (–3); then add it to the second line.

(c) To divide the first line by 2 and multiply it by (–3); then subtract it from the second line.

(d) Both alternatives (a) and (b).

Solution 1:
Gauss elimination aims at isolating variables by multiplying a line by a scalar and subtracting or adding to other lines. In this problem with only two variables, in the final step (as presented in relation 2.5), one variable is multiplied by zero and the other has a non-zero coefficient of 1.

- To do that, the first line can be multiplied by 3/2 and then subtracted from the second line to isolate variable c. This shows that option (a) is correct.

- Note that this is equivalent to multiplying the first line by –3/2 and adding it to the second line, which means (b) is also true. For that, option (d) is the correct answer: $w - 3v/2 = w + (-3v/2)$.

Option (c) does not make any of the coefficients equal to 0 and, as such, it does not work when applying Gauss elimination.

2.3 GAUSS ELIMINATION WITH THE AUGMENTED MATRIX

The Gauss elimination method can be applied not only directly to the system of equations, but also when the linear system is expressed as an augmented matrix. Consider the square matrix **A**, with equal numbers of rows and columns, and let the system coefficients be the elements of **A**:

$$\mathbf{A} = \begin{bmatrix} 2 & 1 \\ 2 & 2 \end{bmatrix}$$

The augmented matrix is represented as the square matrix, **A**, along with an extension of the column vector, **b**, whose elements correspond to the scalars on the right-hand side (RHS) of the equalities:

$$\mathbf{b} = \begin{bmatrix} 6007 \\ 8008 \end{bmatrix}$$

The augmented matrix is then written as [**A**|**b**], and the aim is to find the column vector x such that **Ax** = **b**. Treating the problem in matrix form allows for the resolution of more complex problems, and also promotes computing. The augmented matrix for the system at hand is thus

$$\begin{bmatrix} 2 & 1 & 6007 \\ 2 & 2 & 8008 \end{bmatrix}$$

When the Gauss elimination method is applied to the augmented matrix, the aim is to make linear transformations to the rows of the matrix in such a way the identity matrix **I** is obtained in the first member of the system. Remember that the identity matrix **I** is a square matrix where all the elements are equal to zero except for the main diagonal, which is equal to 1. For example, the 2 × 2 identity matrix **I** is written as

$$\mathbf{I} = \begin{bmatrix} 1 & 0 \\ 0 & 1 \end{bmatrix}$$

In this case, by subtracting the first row from the second row, we obtain

$$\begin{bmatrix} 2 & 1 & 6007 \\ (2-2) & (2-1) & (8008-6007) \end{bmatrix} \leftrightarrow \begin{bmatrix} 2 & 1 & 6007 \\ 0 & 1 & 2001 \end{bmatrix} \tag{2.7}$$

Then, the substitution procedure starts from the bottom. The second row is subtracted from the first row, and then divided over 2, to obtain the identity matrix $\mathbf{I}$ on the left side of the augmented matrix, as follows:

$$\begin{bmatrix} 2 & (1-1) \\ 0 & 1 \end{bmatrix}\begin{vmatrix} (6007-2001) \\ 2001 \end{vmatrix} \leftrightarrow \begin{bmatrix} 2 & 0 \\ 0 & 1 \end{bmatrix}\begin{vmatrix} 4006 \\ 2001 \end{vmatrix} \leftrightarrow \begin{bmatrix} 1 & 0 \\ 0 & 1 \end{bmatrix}\begin{vmatrix} 2003 \\ 2001 \end{vmatrix} \qquad (2.8)$$

Now the solution can be directly read, as the matrix can be written again as a system of equations, obtaining $t = 2003$ and $c = 2001$.

Note that this solution (2003, 2001) is the one single solution for the current linear system, as the restriction lines are crossing each other at one single point, as presented in Figure 2.3.

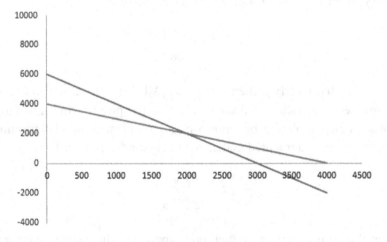

FIGURE 2.3 Single solution at the crossing point (2003, 2001) for the two restriction lines.

QUESTION 2:
Consider the system:

$$\begin{bmatrix} 2 & 1 \\ 4 & 2 \end{bmatrix}\begin{vmatrix} 0 \\ 0 \end{vmatrix}$$

and note that all the RHS parameters are zero.
    For the presented system, classify the following statements (T/F):

(a) The system has the trivial solution (0, 0).
(b) The system has one and only one solution.
(c) The system has an infinite number of solutions.

Solution 2:

Statement (a) is easily proven to be true as the system can be written in equation form, and the solution substituted as follows:

$$\begin{cases} 2x+1y=0 \\ 4x+2y=0 \end{cases} \rightarrow \begin{cases} 2\times0+1\times0=0 \\ 4\times0+2\times0=0 \end{cases}$$

Statement (b) is false, as one can easily come up with an additional solution, a counter-example. For example, in addition to the trivial solution (0, 0), the pair (1, –2) is also a solution:

$$\begin{cases} 2\times1+1\times(-2)=0 \\ 4\times1+2\times(-2)=0 \end{cases}$$

Statement (c) is more difficult to follow; in fact, it is possible to come up with more solutions, but it is needed to show the number of solutions is infinite. Note that both equations in the system represent the same relationship between variables. This can be proven by isolating one of the variables in both equations and observing that the same result is obtained:

$$\begin{cases} 2x+1y=0 \\ 4x+2y=0 \end{cases} \rightarrow \begin{cases} y=-2x \\ 2y=-4x \end{cases} \rightarrow \begin{cases} y=-2x \\ y=-2x \end{cases}$$

This indicates that the system is linearly dependent. Any linear system of equations is said to be linearly dependent if one of the equations can be written as a linear combination of the remaining equations. Because two linearly dependent equations represent the exact same relationship between the variables, one of them can be eliminated without losing any information on the system. As such,

$$\begin{cases} 2x+1y=0 \\ 4x+2y=0 \end{cases} \equiv 2x+1y=0$$

But this system has more variables than equations, and thus is an under-determined system. Note that for any value of x, a value of y can be obtained, and vice versa; therefore, the system presents an infinite number of solutions.

This situation is graphically represented by two coincident lines that correspond to the same expression, $y = -2x$, as in Figure 2.4.

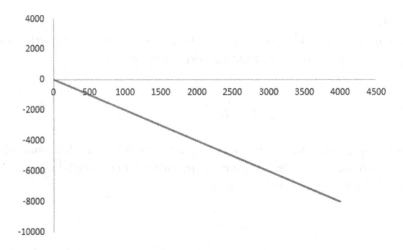

FIGURE 2.4 Infinite solutions for the two coincident restriction lines.

2.4 GAUSS–JORDAN AND THE INVERSE MATRIX

Gauss–Jordan is another matrix-based method, which, in this case, is finding the inverse of the square coefficient matrix A^{-1}. The multiplication of matrix A and its inverse matrix A^{-1} results in the neutral element for matrix multiplication, the identity matrix:

$$A \cdot A^{-1} = A^{-1} \cdot A = I$$

This property is similar for real numbers, where the multiplication of a number (e.g., 2) by its inverse (1/2, that is 2^{-1}) results in 1, the neutral element for numbers multiplication.

Once the inverse matrix has been obtained, A^{-1}, multiplying the inverse matrix by the vector that represents the RHS scalars, b, will result in the system solution's vector, x:

$$A.x = b$$

$$A^{-1}.(A.x) = A^{-1}.b$$

$$(A^{-1}.A).x = A^{-1}.b$$

$$x = A^{-1}.b$$

(2.9)

These procedures are similar to those for solving a real numbers equation, for example: the double of a number is 6, $2x = 6$, with the solution $x = 3$, calculated by dividing 6 by 2. However, dividing by 2 is the same as multiplying by the inverse of 2, which is the opposite number of 2 for multiplication, and

$$2.x = 6$$

$$2^{-1}.(2.x) = 2^{-1}.6$$

$$\left(2^{-1}.2\right).x = 2^{-1}.6$$

$$x = 2^{-1}.6$$

Note that to find the inverse matrix of a square matrix, write the matrix to be inverted on the left side and the identity matrix on the right side; then, make linear transformations to the rows until the identity matrix is obtained in the left square, similarly as presented in relations 2.7 and 2.8.

Let us see how this works on the example. First, the inverse matrix is needed. The first transformation is to subtract the first row from the second row:

$$\begin{bmatrix} 2 & 1 & | & 1 & 0 \\ 2 & 2 & | & 0 & 1 \end{bmatrix} \leftrightarrow \begin{bmatrix} 2 & 1 & | & 1 & 0 \\ (2-2) & (2-1) & | & (0-1) & (1-0) \end{bmatrix} \leftrightarrow \begin{bmatrix} 2 & 1 & | & 1 & 0 \\ 0 & 1 & | & -1 & 1 \end{bmatrix}$$

Then, the transformed second row is subtracted from the first row, and all of the first row is divided over 2:

$$\begin{bmatrix} (2-0) & (1-1) & | & (1-(-1)) & (0-1) \\ 0 & 1 & | & -1 & 1 \end{bmatrix} \leftrightarrow \begin{bmatrix} 2 & 0 & | & 2 & -1 \\ 0 & 1 & | & -1 & 1 \end{bmatrix} \leftrightarrow \begin{bmatrix} 1 & 0 & | & 1 & -1/2 \\ 0 & 1 & | & -1 & 1 \end{bmatrix}$$

Now, the inverse matrix $\mathbf{A}^{-1}$ is obtained on the right side of the augmented matrix:

$$\mathbf{A}^{-1} = \begin{bmatrix} 1 & -1/2 \\ -1 & 1 \end{bmatrix}$$

The prior calculation can be rapidly verified, because the multiplication of a square matrix $\mathbf{A}$ and its inverse matrix $\mathbf{A}^{-1}$ gives the identity matrix:

$$A \cdot A^{-1} = \begin{bmatrix} 2 & 1 \\ 2 & 2 \end{bmatrix} \cdot \begin{bmatrix} 1 & -1/2 \\ -1 & 1 \end{bmatrix} = \begin{bmatrix} 1 & 0 \\ 0 & 1 \end{bmatrix}$$

Finally, multiplying the inverse matrix A^{-1} by the vector b results in the vector solution, x, as follows:

$$x = A^{-1} \cdot b = \begin{bmatrix} 1 \times 6007 + (-1) \times 8008/2 \\ (-1) \times 6007 + 1 \times 8008 \end{bmatrix} = \begin{bmatrix} 6007 - 4004 \\ -6007 + 8008 \end{bmatrix} = \begin{bmatrix} 2003 \\ 2001 \end{bmatrix} \quad (2.10)$$

With this matrix multiplication, the solution is obtained: the number of tables is 2003 and the number of chairs is 2001.

QUESTION 3:

Revisiting the inverse matrix:

$$A^{-1} = \begin{bmatrix} 1 & -1/2 \\ -1 & 1 \end{bmatrix}$$

From the inverse matrix, if the first parameter in b increases by 1...

(a) How much does the value of the first variable grow?

(b) How much does the value of the second variable grow?

Solution 3:

Suppose any vector b represents the RHS of a system of two linear equations:

$$b = \begin{bmatrix} b_1 \\ b_2 \end{bmatrix}$$

Then, by multiplying the inverse matrix A^{-1} by b, we obtain the following:

$$A^{-1} \cdot b = \begin{bmatrix} 1 & -1/2 \\ -1 & 1 \end{bmatrix} \cdot \begin{bmatrix} b_1 \\ b_2 \end{bmatrix} = \begin{bmatrix} 1 \cdot b_1 & -b_2/2 \\ -1 \cdot b_1 & 1 \cdot b_2 \end{bmatrix}$$

By this means, if the first RHS parameter, b_1, increases by 1, then the value for the first variable also increases by 1, while the value for the second variable decreases by 1 (its alteration is -1).

A similar rationale can be made for a unit alteration of the second RHS parameter, b_2, and the new solution can be directly calculated without solving the linear system from the starting point. For that, use the previous solution and sum the alterations estimated from the inverse matrix coefficients. For example, using the current instance in relation 2.10:

- Assuming b_1 increases by 1 to 6008, then the first variable t also increases by 1 to 2004, while the second variable c decreases by 1 to 2000.

  Note that this result is confirming the prior reasoning of adding one big-blue piece, where one more table is built and one chair is undone.

- After that alteration, suppose b_2 increases by 1 to 8009; then the first variable t diminishes by half to 2003.5, while the second variable c increases by 1 to 2001.

  Note that this result is related to the prior reasoning of adding two small-red pieces, then two more chairs are built while one table is undone.

2.5 CRAMER'S RULE AND DETERMINANTS

Cramer's rule is a direct formula for solving systems of linear equations based on the computation of determinants. Using relation 2.11, Cramer's rule finds the value for each component x_i of the solution vector $\mathbf{x}$:

$$x_i = \frac{\det(\mathbf{A}_i)}{\det(\mathbf{A})} \tag{2.11}$$

where $\mathbf{A}_i$ is the matrix that results from substituting the i-th column of matrix $\mathbf{A}$ with the vector $\mathbf{b}$, such that $\mathbf{A}\mathbf{x} = \mathbf{b}$.

In the example, the following two formulae to solve the problem are obtained:

$$\begin{cases} t = \dfrac{\begin{vmatrix} 6007 & 1 \\ 8008 & 2 \end{vmatrix}}{\begin{vmatrix} 2 & 1 \\ 2 & 2 \end{vmatrix}} \\[4ex] c = \dfrac{\begin{vmatrix} 2 & 6007 \\ 2 & 8008 \end{vmatrix}}{\begin{vmatrix} 2 & 1 \\ 2 & 2 \end{vmatrix}} \end{cases}$$

Now, it is only needed to compute three determinants to obtain the solution. The determinant of any square matrix can be computed using Laplace's expansion formula:

If **A** is a square ($n \times n$)-sized matrix, with coefficients a_{ij} for every row i and column j, then the determinant of **A**, det(**A**) or $|A|$, can be computed as a weighted sum of the determinants of n sub-matrices of A, each $(n - 1) \times (n - 1)$ in size. The determinant of **A** is computed using relation 2.12 for any row i or column j (the result is independent of the row or column chosen).

$$|A| = \sum_{j'=1}^{n} \alpha_{ij'} (-1)^{i+j'} |M_{ij'}| = \sum_{i'=1}^{n} \alpha_{i'j} (-1)^{i'+j} |M_{i'j}| \qquad (2.12)$$

where $|M_{ij}|$ is the $(n - 1) \times (n - 1)$ matrix that results from eliminating row i and column j from matrix **A**.

For a 2×2 matrix, this computation is simple enough:

$$\begin{vmatrix} a_{11} & a_{12} \\ a_{21} & a_{22} \end{vmatrix} = a_{11} \cdot a_{22} - a_{12} \cdot a_{21}$$

Applied to the problem at hand, then

$$\begin{cases} t = \dfrac{\begin{vmatrix} 6007 & 1 \\ 8008 & 2 \end{vmatrix}}{\begin{vmatrix} 2 & 1 \\ 2 & 2 \end{vmatrix}} = \dfrac{6007 \times 2 - 8008 \times 1}{2 \times 2 - 2 \times 1} = \dfrac{12014 - 8008}{4 - 2} = \dfrac{4006}{2} = 2003 \\[3em] c = \dfrac{\begin{vmatrix} 2 & 6007 \\ 2 & 8008 \end{vmatrix}}{\begin{vmatrix} 2 & 1 \\ 2 & 2 \end{vmatrix}} = \dfrac{8008 \times 2 - 6007 \times 2}{2 \times 2 - 2 \times 1} = \dfrac{16016 - 12014}{4 - 2} = \dfrac{4002}{2} = 2001 \end{cases}$$

Note that when a determinant of a matrix is equal to zero, this indicates that the matrix is linearly dependent. Because of this, it can be stated that for 2×2 systems:

- If all the determinants are non-zero, then the system has one single solution.

This case is represented by two crossing lines at one single point, as shown in Figure 2.3.

- If the determinant of the coefficient matrix **A** is zero, then this determinant is located in the fraction denominators.

  o If the numerator's determinant is also zero, then the system is indeterminate, and thus has an infinite number of solutions.

  Note that this case is represented by two coincident lines (e.g., Figure 2.4).

  o If the numerator's determinant is not zero, then the system is incompatible and thus no solution exists.

QUESTION 4:

For the linear system:

$$\begin{cases} 2t + 1c = 6007 \\ 2t + 1c = 8008 \end{cases}$$

Applying Cramer's rule to the presented system of equations, which of the following statements is true?

(a) $t = 2003; c = 2002$.

(b) $t = 2004; c = 2001$.

(c) The system has no solution.

(d) None of the above.

Solution 4:

Let us write this problem in order to solve it using Cramer's rule:

$$t = \frac{\begin{vmatrix} 6007 & 1 \\ 8008 & 1 \end{vmatrix}}{\begin{vmatrix} 2 & 1 \\ 2 & 1 \end{vmatrix}} = \frac{6007 \times 1 - 8008 \times 1}{2 \times 2 - 2 \times 2} = \frac{-2001}{0}, \; n.d$$

$$c = \frac{\begin{vmatrix} 2 & 6007 \\ 2 & 8008 \end{vmatrix}}{\begin{vmatrix} 2 & 1 \\ 2 & 1 \end{vmatrix}} = \frac{8008 \times 2 - 6007 \times 2}{2 \times 2 - 2 \times 2} = \frac{4002}{0}, \; n.d$$

The denominator's determinant is zero, but the determinants in the numerators are not zero; this occurs with an incompatible system, for which no solution exists. Thus, option (c) is correct.

- In fact, the first member of both equations is the same; however, the sum result in the RHS is either 6007 or 8008, but it can never take the two values at the same time.

- A graph representation of the two lines presents two parallel lines with the same slope, but one line crosses the vertical axis at point 6007 and the other at point 8008; these parallel lines never cross each other, also indicating that the system has no solution (Figure 2.5).

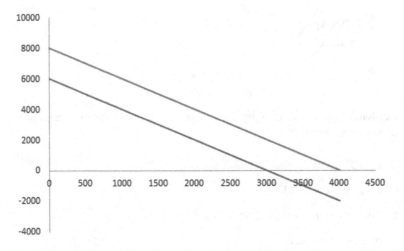

FIGURE 2.5   No solution for the two parallel restriction lines.

Cramer's rule is commonly applied for very small linear systems, given its very simple and direct formulation; however, computational issues and cancellation errors make it very inefficient for a large matrix, and thus for large linear systems.

2.6 CONCLUDING REMARKS

LA is an important topic concerning optimal decision-making, given the simple solution methods that can efficiently solve problems based on systems of equations. However, LA presents important shortcomings when applied in the context of optimal decision-making, namely:

- Some solutions that do not represent real possibilities (such as non-integer or non-positive values) are feasible solutions when only applying LA-based decision-making tools.

- LA methods consider the simultaneous solution of all the equalities for a linear system, thus no slack is allowed; that is, 100% of the resources or components must be used; if this is not the case, the solutions obtained with LA-based methods might be suboptimal.

- There can be one of three types of solutions for a linear system of equations: (i) one and only one feasible solution; (ii) an infinite number of solutions; or (iii) feasible solutions do not exist.

- However, when such a solution does exist, LA methods obtain one single solution; this attribute does not allow any degree of freedom for improved decision-making or the introduction of new variables within the decision maker's point of view.

For these reasons, more sophisticated tools that cover these shortcomings and allow for better decision-making are needed. One such tool is presented in Chapter 3, when we introduce linear programming (LP) that supports the decision maker in the optimal selection of one alternative using one single attribute.

Linear Programming Basics

THIS CHAPTER ADDRESSES THE basics of linear programming (LP), from the simple graphical approach to the algebraic form, the LP Simplex tableau, and the matrix form with the updating procedure for the inverse matrix. In this chapter, the prior linear algebra (LA) approach is enlarged, and the LP Simplex method addresses a linear system of equations at each iteration; new attributes that improve decision-making are obtained, being directly related with slack variables, dual values, and sensitive parameters.

3.1 INTRODUCTION

The typical LP application addresses the rational allocation of limited resources to different activities; these activities compete with each other to obtain the best allocation of resources and the optimal solution, which is evaluated to drive the optimum value.

Let it be the furniture factory problem, and an instance presenting friendly parameters for easier calculations:

A furniture factory builds tables (t) at a profit of 40 euros per table, and chairs (c) at a profit of 30 euros per chair (Figure 3.1).

DOI: 10.1201/9781315200323-3

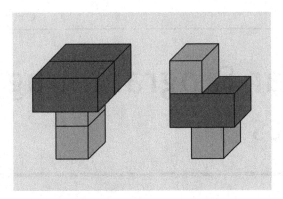

FIGURE 3.1   The furniture factory problem: table and chair.

Suppose that

- only 80 small-red pieces and 60 big-blue pieces are available; and
- the upper capacity for the chairs' assembly line is 30.

What combination of tables and chairs would maximize the total profit?

The LP mathematical models only use linear functions to formulate both the objective function and the restrictions. Thus, it is necessary to properly define the decision variables and then to formulate the linear relations in the model.

The *decision variables* are the alternatives or options the decision maker may consider. In this problem, the alternatives are related to both the quantity of tables to produce, t, and the quantity of chairs, c. Note that the decision maker may opt either to produce only tables, to produce only chairs, or to produce a mix of tables and chairs. The mix and the related quantities of tables and chairs are defined in such a way as to optimize the objective function and satisfy the restrictions set.

In fact, the LP model for the furniture factory problem considers:

- **Objective function:** Linear function to be optimized (maximized or minimized), as it represents the system performance; in the furniture problem:

$$\max Z = 40t + 30c$$

- **Restrictions:** Linear constraints to be satisfied by the optimal solution; in the problem at hand:

$$\begin{cases} 2t + 1c \leq 60 \\ 2t + 2c \leq 80 \\ 1c \leq 30 \end{cases}$$

- **Non-negativity conditions:** The variables must be positive or zero:

$$t, c \geq 0$$

Note that the common LP model considers the maximization of a linear objective function that will satisfy a set of *less than or equal to* restrictions. A simple and general explanation by contradiction follows:

- If a set of *more than or equal to* restrictions occur, the variables increase without upper bounds and the objective function's value increases without any bound, then the maximum value tends to infinity, and the model is not consistent; for that, a consistent restrictions set for a maximization LP model is the *less than or equal to* type.

$$\max [Z] = \mathbf{c.x}$$

subject to

$$\mathbf{A.x} \leq \mathbf{b}$$

$$\mathbf{x} \geq \mathbf{0}$$

It is common for LP models to also consider the minimization of a linear objective function that will satisfy a set of *more than or equal to* restrictions. Again, by contradiction:

- If a set of *less than or equal to* restrictions occur, the variables diminish without lower bounds, namely until 0, since non-negativity still applies; then the objective function's value diminishes without any lower bound, tending to 0 within the associated trivial solution (all variables with 0 value).

$$\min [Z] = \mathbf{c.x}$$

subject to

$$\mathbf{A.x} \geq \mathbf{b}$$

$$\mathbf{x} \geq 0$$

For the furniture problem at hand, the LP model presents the maximization form:

$$\max Z = 40t + 30c$$

subject to

$$\begin{cases} 2t + 1c \leq 60 \\ 2t + 2c \leq 80 \\ 1c \leq 30 \end{cases} \qquad (3.1)$$

$$t, c \geq 0$$

The LP model assumes several properties:

- **Proportionality:** The contribution of each activity to the objective and the restrictions is proportional to the level of the activity; namely:

  o By producing one table, the gain is 40 and two big-blue pieces are used; then, to produce two tables, the gain is 80 and four big-blue pieces are used; to produce three tables, the gain is 120 and six big-blue pieces are used, and so on.

  o By producing one chair, the gain is 20 and two small-red pieces are used; then, to produce two chairs, the gain is 60 and four small-red pieces are used; to produce three chairs, the gain is 90 and six small-red pieces are used, and so on.

- **Additivity:** The global value for the objective or the utilization of resources results from the addition of the individual contribution of all the activities, namely:

  o By producing one table and one chair, the total gain is 70, by adding the table's contribution to the profit, 40, to the chair's contribution,

30; note that producing two tables and two chairs, the total gain is 140; and the gain of three tables and three chairs is 210.

o Concerning the consumption of big-blue pieces, by producing one table and one chair, the total consumption is 3, by adding the table's contribution to the resource's consumption, 2, to the chair's contribution, 1; then, to produce two tables and two chairs, the total consumption is 6, by adding the table's contribution, 4, to the chair's contribution, 2; and the total consumption of 20 tables and 20 chairs is 60 big-blue pieces.

o Concerning the consumption of small-red pieces, by producing one table and one chair, the total consumption is 4, by adding the table's contribution to the resource's consumption, 2, to the chair's contribution, 2; then, to produce two tables and two chairs, the total consumption is 8, by adding the table's contribution, 4, to the chair's contribution, 4; and the total consumption of 20 tables and 20 chairs is 80 small-red pieces.

- **Continuity or divisibility:** The decision variables can present non-integer values, e.g., the variables can be considered continuous, e.g., as real numbers and non-negative variables. For example:

  o By producing 1/2 a table, the gain would be only 20 and one big-blue piece is utilized; also, to produce 1/4 of a table, the gain would be only 10 and 1/2 a big-blue piece is needed.

  o However, the feasibility of such a continuous variable, t, will require confirmation in the real world for the possible implementation of non-integer solutions.

  o In the same way, for the continuous variable, c, by producing 1/2 a chair, the gain would be only 15 and one small-red piece is utilized; also, to produce 1/4 of a chair, the gain would be only 7.5 and 1/2 a small-red piece is needed.

3.2 GRAPHICAL APPROACH

The LP model for the furniture factory problem presents only two decision variables, t and c, and the linear functions can be manipulated to express one variable in relation to the other, for example, $c(t)$. In addition, due to the non-negativity property of both variables, the straight lines within the

LP model are only addressed in the positive part of the plane, *Otc*, that is, the first quarter in the plane *Oxy*.

For each restriction, the related straight line bounds the feasible region, which includes the points of coordinates (t, c) that satisfy all restrictions. Namely,

- Concerning the first restriction on the availability of big-blue pieces:

$$2t + 1c \leq 60$$

$$c \leq 60 - 2t$$

the feasible region considers the area below the straight line that crosses the vertical axis at 60 (when $t = 0$, then $c = 60$), with slope –2, and then crosses the horizontal axis at 30 (when $c = 0$, then $t = 30$), as indicated by the darker line in Figure 3.2a.

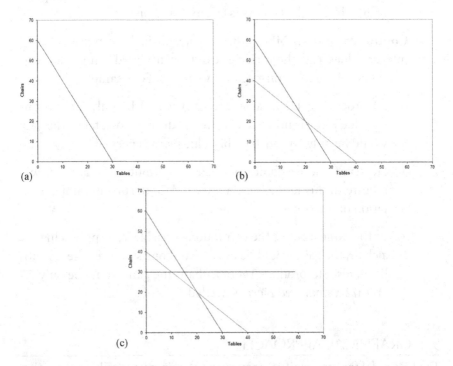

FIGURE 3.2 Graphical approach to LP. (a) Adding the first resource's restriction, for big-blue pieces. (b) Adding the second resource's restriction, for small-red pieces. (c) Adding the third resource's restriction, for chairs' assembly line.

- Concerning the second restriction on the availability of small-red pieces:

$$2t + 2c \leq 80$$

$$2c \leq 80 - 2t$$

$$c \leq 40 - t$$

the feasible region considers the area below the straight line that crosses the vertical axis at 40 (when $t = 0$, then $c = 40$), with slope -1, and then crosses the horizontal axis at 40 (when $c = 0$, then $t = 40$), as indicated by the light gray line in Figure 3.2b.

- Concerning the third restriction on the upper capacity for the chairs' assembly line:

$$c \leq 30$$

the feasible region considers the area below the straight and horizontal (with slope 0) line that crosses the vertical axis at 30, as indicated by the black line in Figure 3.2c.

Some concepts of interest are now introduced, noting that they will be useful later on.

- Edge: The line segment that bounds the feasible region; e.g., the effective part of the straight lines $c = 60 - 2t$; $c = 40 - t$; $c = 30$.

- Feasible solution: The solution that satisfies all restrictions; e.g., the solutions with coordinates $(0, 0)$; $(30, 0)$; $(0, 30)$.

- Feasible region: The region with all feasible solutions; that is, a bounded search region where feasible solutions can be found (see gray background in Figure 3.3).

- Infeasible solution: The solution that does not satisfy at least one of the restrictions; e.g., the solution $(0, 40)$ does not satisfy the third restriction on the chairs' assembly line; and the solution $(40, 0)$ does not satisfy the first restriction on the big-blue pieces.

- Optimal solution: The feasible solution corresponding to the best value ("optimal") of the objective function.

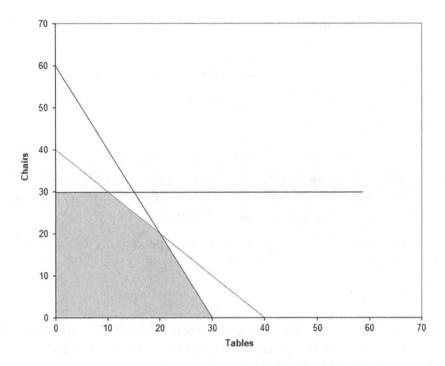

FIGURE 3.3   The feasible region, in gray background.

In order to find the optimal solution, let us focus on the objective function with value Z for the dependent variable:

$$Z = 40t + 30c$$

When specifying a constant value Z, all the points defined by the above equation and represented by a straight line where all points have the same value, Z: an isoline. Then, assuming a constant value Z, and manipulating the objective function in order to variable c, the relation for the set of objective function isolines is obtained:

$$30c = Z - 40t$$

$$c = \frac{Z}{30} - \frac{40}{30}t$$

Given Z, the objective function's isoline represents all those points in the straight line with slope $(-4/3)$ that crosses the vertical axis at point $Z/30$, as shown in Figure 3.4. Note that

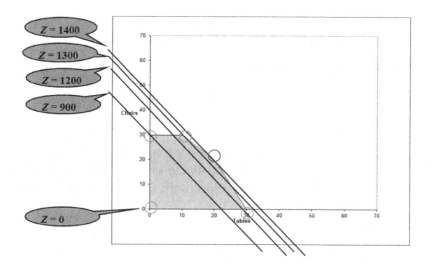

FIGURE 3.4 Parallel isolines improve the objective function's value Z.

- When all the isolines for the objective function present the same slope (–4/3), then they are all parallel straight lines.

- When the value Z successively increases, the isolines cross the vertical axis in successively higher points, Z/30; then, the related straight lines are increasingly higher too.

- The procedure for increasing Z will terminate when the isoline reaches the last point in the feasible region, that is, the last vertex or corner point (in darker circle, Figure 3.4).

- In fact, additional increases in the objective function value, Z, will drive the isolines completely out of the feasible region, without reaching any feasible point.

Typically, the LP problem presents a feasible region that is edge-bounded, with several vertexes or corner points, and an optimal solution.

- The best corner point of the feasible region must be an optimal solution; as in Figure 3.4, the corner points are associated with the values 0, 900, 1200, 1300, and finally the optimal solution (20, 20)* for Z* = 1400.

- If the optimal solution is one and only one, then it is a feasible corner point; again, additional increases in Z will drive the isoline out of the feasible region.

- If the optimal solution is multiple, it considers two corner points and the edge, that is, the line segment between them.

  This situation could occur when the slope of the objective function equals the slope of a restriction; note that the objective function's slope (–4/3) is the negative ratio between the profits for tables, 40, and chairs, 30; then:

  o The first restriction, on the big-blue resources, has a slope –2; this ratio could be matched by (–40/20) or (–60/30), with the profits for tables and chairs, respectively, in the numerator and the denominator.

  o The second restriction, on the small-red resources, has a slope –1; this ratio could be matched by (–40/40) or (–30/30), in a similar manner.

- Multiple optimal solutions: The various and infinite solutions that present the same optimal value of the objective function.

  This situation would include the two corner points and the edge between them; the edge at hand and the objective function would present the same slope, –2 or –1, as discussed in the prior bullet point.

- Adjacent solutions: The pair of feasible corner points that share the same restriction line; that is, these vertexes are connected by a line segment that represents a linear restriction.

For an LP problem, it is observed that the optimal solution is a feasible corner point; it can be reached from an adjacent corner point, which can be directed from another corner, and so on, since it is known that the optimal solution cannot be an interior point.

The LP Simplex also includes these simple guidelines, only focusing on the solutions associated with feasible corner points; then, the corresponding graphical approach for the furniture factory problem follows:

- Since it is a feasible point, the initial guess is the corner (0, 0); then the initial Z value for the objective function is 0 too.

- The method searches to improve the current solution by evaluating the adjacent corners, (30, 0) and (0, 30); the associated values for the objective function are, respectively, 1200 and 900.

- The edge selection aims to improve the objective function the most; thus, the former corner is selected, (30, 0), improving the objective function by 40 per table, and adding 1200 to the current Z value.

   o Note that the other corner point (0, 30) is associated with a lower value, $Z = 900$, with the objective function improving by only 30 per chair.

- The optimality test verifies if some of the edges that connect the current corner point (30, 0) may improve the objective function; now, only the corner (20, 20) requires attention, with a total value of $Z = 1400$.

- The corner point (20, 20)* with a value of $Z* = 1400$ is confirmed as the optimal solution, since the next adjacent corner (10, 30) is associated with a reduction to 1300.

The Simplex method only evaluates the objective function in the corner points; the feasible corner corresponding to the objective function's highest value is the optimal corner point, as presented in Figure 3.5.

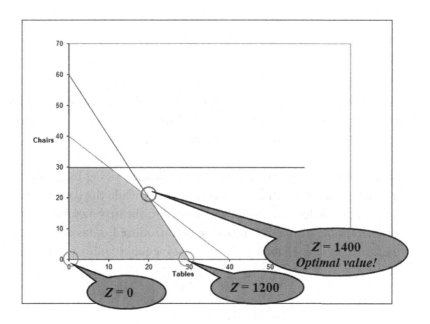

FIGURE 3.5   Improving the objective function's value with selected corners.

The LP Simplex is a very efficient method, as detailed in the following sections. The presented geometric representation corresponds to simple algebraic procedures.

3.3 ALGEBRAIC FORM

In this section, the algebraic form for the LP Simplex method is presented, and the relations with the geometric representation of lines, corner points, and regions are revisited. In addition, the usual LP terminology is introduced for a better understanding of the reasoning behind the procedures.

The LP problem with n-variables and m-restrictions is represented in algebraic form as a linear system of equations:

$$\min Z = c_1 x_1 + c_2 x_2 + \cdots + c_n x_n$$

subject to

$$\begin{cases} a_{11}x_1 + a_{12}x_2 + \cdots + a_{1n}x_n = b_1 \\ a_{21}x_1 + a_{22}x_2 + \cdots + a_{2n}x_n = b_2 \\ (...) \\ a_{m1}x_1 + a_{m2}x_2 + \cdots + a_{mn}x_n = b_m \end{cases}$$

$$x_i \geq 0, i = 1..n$$

The restrictions set for the furniture factory's problem is a set of inequalities from the *less than or equal to* type, and the slack variable, s_i, is introduced to transform them into a set of equalities. Namely,

- Concerning the first restriction, the availability of big-blue pieces is presented on the right-hand side (RHS), 60, while the consumption to produce tables and chairs is presented in the first member; let the slack variable to the first resource be introduced, s_1, as an LP non-negative variable that converts the first inequality into an equation:

$$2t + 1c \quad \leq 60$$

$$2t + 1c + 1s_1 = 60$$

o When the consumption of big-blue pieces is 60, it is equal to the related availability and there is no slack to this first resource, then $s_1 = 0$; the first resource is exhausted and the first restriction is said to be active, since it is actively constraining the furniture's production, and the problem solution too.

o When the consumption is 59, 58, 57,..., 50, then the slack variable s_1 takes 1, 2, 3,..., 10, respectively; the first restriction is said to be not active, since the solution is not being constrained by the first resource.

o Could the consumption be 61, 62, 63,..., 70? No! Firstly, this is not possible in a real situation, all these big-blue pieces are not in stock. Secondly, the slack variable s_1 would take the negative values –1, –2, –3,..., –10, and the LP solution would not be feasible.

• Concerning the second restriction, the availability of small-red pieces is presented on the RHS, 80, while the consumption to produce tables and chairs is presented in the first member; let the slack variable to the second resource be introduced, s_2, as an LP non-negative variable that converts the second inequality into an equation:

$$2t + 2c \leq 80$$

$$2t + 2c + 1s_2 = 80$$

o When the consumption of small-red pieces is 80, it is equal to the associated availability and there is no slack to the second resource, then $s_2 = 0$; the second resource is exhausted and the second restriction is said to be active.

o When the consumption is 79, 78,..., 70, then the slack s_2 takes the values 1, 2,..., 10, respectively; thus, the second restriction is not active.

o Could the consumption be 81, 82,..., 90? No, again! All these small-red pieces would not be available in stock, and the slack variable s_2 would take the negative values –1, –2,..., –10, respectively.

- As for the third restriction, the capacity's upper limit in the chairs assembly is presented on the RHS, 30, while the chairs production is presented in the first member; let the non-negative slack variable be introduced, s_3, and convert the third inequality into an equation:

$$c \quad \leq 30$$

$$c + 1s_3 = 30$$

- o When the chairs production is 30, there is no slack to the third resource, and $s_3 = 0$; the third resource is exhausted and the third restriction is active.

- o When the chairs production is 29, 28,…, 20, then the slack s_3 takes the values 1, 2,…, 10, respectively; thus, the third restriction is not active.

- o Could the chairs production be 31, 32,…, 40? No, once again! The assembly's capacity would not reach that value; additionally, the slack variable s_3 would take negative values of –1, –2,…, –10.

Slack variables are not contributing to profit maximization, therefore all the related coefficients in the objective function are made 0. The LP model is thus assuming the algebraic form of a linear function to maximize and a set of linear equations to satisfy:

$$\max Z = 40t + 30c + 0s_1 + 0s_2 + 0s_3$$

subject to $\hfill$ (3.2)

$$\begin{cases} 2t + 1c + 1s_1 & = 60 \\ 2t + 2c \quad\;\; + 1s_2 & = 80 \\ \quad\;\, 1c \quad\quad\;\; + 1s_3 & = 30 \end{cases}$$

From the LA basics, it should be noted that

- The linear system of equations presents one and only one solution when the number of variables, n, is equal to the number of linearly independent equations, m; in this case, the solution is unique but it does not present any degree of freedom, and the decision maker is faced with one single alternative, a unique choice.

- If $n < m$, the system is non-tractable; no feasible solutions exist, as it is an overdetermined system.

- If $n > m$, the system presents infinite solutions; in this case, the decision maker is faced with many degrees of freedom, by selecting the values for a set of $(n - m)$ variables, and then calculating the other m-variables.

- For that, the LP optimization problem typically occurs when $n > m$.

In this way, the restrictions set in Section 3.2 correspond to a linear system of equations with five variables (t, c, s_1, s_2, s_3), and three equations; this system requires to first define two values, and only then the other three variables can be calculated from the three equations, within the LA's single solution framework.

$$\begin{cases} 2t + 1c + 1s_1 & = 60 \\ 2t + 2c \quad + 1s_2 & = 80 \\ \quad 1c \quad + 1s_3 = 30 \end{cases}$$

Some useful terminology is now introduced:

- Non-basic variables: The $(n - m)$ variables are estimated as zero value; in this problem, two non-basic variables are made 0, allowing one single solution to be found for the system.

- Basic variables: The other m-variables can assume non-zero values; now, the single solution for the system with three linear equations presents the values associated with the other three variables.

- Basic solution: The non-negative values are obtained for the basic variables, after the non-basic variables are made zero.

- Degenerate solution: In the case that (at least) one of the basic variables presents zero value(s), then such a solution would be considered "degenerate".

After introducing slack variables, an iterative approach based on LA is used to address the LP in algebraic form. For simplicity, a trial-and-error method starts with the origin point $(0, 0)$ as the initial guess. In fact, when the production of tables and chairs is set to zero, $t = c = 0$, all the slack variables assume the values of the resources' availability. Respectively, 60, 80, and 30 for the first iteration.

3.3.1 First Iteration

Assuming:

$$\begin{cases} t = 0 \\ c = 0 \end{cases} \Rightarrow \begin{cases} s_1 = 60 \\ s_2 = 80 \text{ then } Z = 0 \\ s_3 = 30 \end{cases}$$

This solution can be improved by 40 for each unit of variable t, that is 40 euros per table, if this variable is greater than zero; then, variable t enters the solution basis.

The highest value for variable t is obtained from the minimum ratio when the first resource (big-blue pieces) is exhausted; then, the slack variable s_1 is reduced to 0 (null, zero value) and it exits the solution basis:

$$\begin{cases} \theta_1 = {}^{60}\!/_2 = 30 \\ \theta_2 = {}^{80}\!/_2 = 40 \\ \theta_3 = {}^{30}\!/_0 \ n.d. \end{cases}$$

- Comparing with the graphic representation in Figure 3.5, when starting from the origin $(0, 0)$, these procedures correspond to searching the next corner point by selecting the horizontal direction instead of the vertical direction, and then continuing in this direction until the feasible region ends, at the adjacent corner $(30, 0)$ with a value of 1200.

By Gauss elimination, addressing the new basic variable, t, the next solution will present a value that is improved by

$$\Delta Z = c_1 . \theta_1$$

$$= 40 \times 30 = 1200$$

For that, the Gaussian procedures consider relation 3.2:

- Dividing line one by 2.
- Subtracting line one directly from line two.
- Multiplying line one by 20 and subtracting from the objective line.

$$\max Z = 0t + 10c - 20s_1 + 0s_2 + 0s_3$$

subject to (3.3)

$$\begin{cases} 1t + 0.5c + 0.5s_1 & = 30 \\ 0t + 1c - 1s_1 + 1s_2 & = 20 \\ 1c & + 1s_3 = 30 \end{cases}$$

3.3.2 Second Iteration

Assuming

$$\begin{cases} s_1 = 0 \\ c = 0 \end{cases} \Rightarrow \begin{cases} t = 30 \\ s_2 = 20 \quad \text{then} \quad Z = 1200 \\ s_3 = 30 \end{cases}$$

This solution can be improved by 10 euros for each unit of variable c, that is, if this variable is greater than zero; therefore, variable c enters the solution basis.

The highest value for c is obtained from the minimum ratio; then, the second resource (small-red pieces) is exhausted; for that, the slack variable, s_2, exits the basis, since it has been reduced to zero.

$$\begin{cases} \theta_1 = {}^{30}\!/_{0.5} = 60 \\ \theta_2 = {}^{20}\!/_1 = 20 \\ \theta_3 = {}^{30}\!/_1 = 30 \end{cases}$$

- Comparing with the graphic representation in Figure 3.5, when starting from the corner (30, 0), these procedures correspond to searching the adjacent corner point by selecting the ramping edge and continuing until the feasible region ends, at the adjacent corner (20, 20) with a value of 1400.

Addressing the new basic variable, c, the next solution will be improved by

$$\Delta Z = c_2 \cdot \theta_2$$

$$= 10 \times 20 = 200$$

Applying Gauss elimination to relation 3.3:

- Dividing line two by the pivot, 1; this step is trivial, it can be skipped.
- Multiplying line two by 1/2 (i.e., dividing by 2), and subtracting to line one.
- Subtracting line two directly from line three.
- Multiplying line two by 10 and subtracting from the objective line.

$$\max Z = 0t + 0c - 10s_1 - 10s_2 + 0s_3$$

subject to $\hspace{6cm}$ (3.4)

$$\begin{cases} 1t + 0c + 1s_1 - 0.5s_2 & = 20 \\ 1c - 1s_1 + 1s_2 & = 20 \\ 0c + 1s_1 - 1s_1 + 1s_3 = 10 \end{cases}$$

3.3.3 Third Iteration

Assuming

$$\begin{cases} s_1 = 0 \\ s_2 = 0 \end{cases} \Rightarrow \begin{cases} t = 20 \\ c = 20 \\ s_3 = 10 \end{cases} \text{ then } Z = 1400$$

This solution cannot be further improved because of the negative coefficients on the objective function. This is the optimal solution!

• Comparing with the graphic representation in Figure 3.5, when located at corner (20, 20) with a value of 1400, it is not possible to find an adjacent corner point where the objective function would increase.

To maximize the objective function while satisfying a set of *less than or equal to* restrictions, the LP Simplex method can be summarized in the following steps:

0) Previous step

Set the LP in algebraic form by introducing slack variables to obtain linear equations.

1) Initialization

Obtain a first basic solution by assuming zero for the $(n - m)$ values of the non-basic variables; this is a feasible solution since it satisfies all the restrictions within the linear system of equations.

2) Optimality test

Verify if the present basic solution is optimal, checking if all the objective function coefficients in terms of the non-basic variables are negative. Otherwise, continue the optimum search within the iterative cycle in step 3.

3) Iterative cycle

a) **Direction of search:** Select the non-basic variable that is entering the basic solution, by choosing the positive coefficient of the objective function with the largest value.

b) **Length of movement:** Select the basic variable that is leaving the solution by the minimum ratio test, by choosing the variable that first reaches zero.

c) **Obtaining the new basic solution:** Use Gauss elimination procedures to obtain the identity vector in the column corresponding to the new basic variable.

For minimizing a linear objective function while satisfying a set of *greater than or equal to* restrictions, the referred steps for the LP Simplex method need some adjustments, the most important of which are described below.

- Due to the opposite type of restrictions, typically, excess variables need to be introduced and then subtracted to convert the inequalities set into a system of equations.

- The initialization point cannot be the axis origin, the trivial solution with the decision variables taking a zero value; usually, artificial variables are introduced or other methods are needed.

- The optimality test checks if all the objective function coefficients in terms of the non-basic variables are positive, in such a way that the objective function cannot be improved.

3.4 TABLEAU FORM

In general, following the steps presented in the previous section, the LP Simplex can be developed in tableau form, and the procedures simplified. In fact, the Simplex tableau only saves the main information, namely,

- The coefficients for restrictions, and for the objective function.

- The RHS parameters in the restrictions.

- The basic variable in each equation.

Previous step

The LP is set by introducing slack variables to obtain linear equations and, as in relation 3.2 for the algebraic form, the first tableau is obtained (Table 3.1).

Initialization

Assuming that the non-basic variables take a value of 0, the system solution can be directly read from the columns associated with the identity

TABLE 3.1   First Tableau – Initial Basic Solution with Slack Variables, $Z = 0$.

Basis	t	c	s_1	s_2	s_3	RHS
s_1	2	1	1	0	0	60
s_2	2	2	0	1	0	80
s_3	0	1	0	0	1	30
Z	40	30	0	0	0	0

matrix; that is, the basic variables take the respective values presented on the RHS.

$$\begin{cases} t = 0 \\ c = 0 \end{cases} \Rightarrow \begin{cases} s_1 = 60 \\ s_2 = 80 \\ s_3 = 30 \end{cases} \text{ then } Z = 0$$

Optimality test?
The solution is not optimal because there are positive coefficients for the non-basic variables, 40 and 30, thus the objective function can be improved.

3.4.1 Iterative Cycle: First Iteration

1) **Direction of search:** Variable t, in the first column ($j = 1$), enters the solution basis; it corresponds to the positive coefficient of the objective function with the largest value, 40.

2) **Length of movement:** Applying the minimum ratio test, the select slack variable, s_1, is the exiting variable, since the first line of resources is exhausted ($i = 1$); then, the pivot element is located in the first line and the first column, $a_{11} = 2$ (in light gray background at Table 3.1).

3) **Obtaining the new basic solution:** Using Gauss elimination and an approach similar to the algebraic form (Section 3.2):

   o Dividing line one by 2.

   o Subtracting line one directly from line two.

   o Multiplying line one by 20 and subtracting from the objective line; note that the value for the objective function becomes negative, the tableau is presenting the opposite value, thus the absolute value is to be considered as the real objective function's value.

Performing the referred calculations, as in the algebraic relation 3.3, the second tableau is thus obtained.

3.4.2 Second Tableau

Again, the non-basic variables take a value of 0, and the basic variables take the respective values presented on the RHS (Table 3.2).

TABLE 3.2   Second Tableau – Variable t Entered the Solution Basis, $Z = 1200$.

Basis	t	c	s_1	s_2	s_3	RHS
t	1	0.5	0.5	0	0	30
s_2	0	1	−1	1	0	20
s_3	0	1	0	0	1	30
Z	0	10	−20	0	0	**−1200**

$$\begin{cases} s_1 = 0 \\ c = 0 \end{cases} \Rightarrow \begin{cases} t = 30 \\ s_2 = 20 \quad \text{then} \quad Z = 1200 \\ s_3 = 30 \end{cases}$$

Optimality test?

The solution is still not optimal because there is a positive coefficient, 10, for the non-basic variable, c.

Iterative cycle

1) **Direction of search:** Entering variable c ($j = 2$).

2) **Length of movement:** Apply the minimum ratio test; then the slack variable s_2 ($i = 2$) exits the solution basis, and the pivot element is located in the second line and second column, $a_{22} = 1$ (in light gray background at Table 3.2).

3) **Obtaining the new basic solution:** Using Gauss elimination once again, as occurred in the algebraic form:

   o Dividing line two by the pivot, 1; this step is trivial, it can be skipped.

   o Multiplying line two by 1/2 (i.e., dividing by 2), and subtracting to line one.

   o Subtracting line two directly from line three.

   o Multiplying line two by 10 and subtracting from the objective line; again, a negative value is presented in the tableau, and the opposite value is considered for the objective function.

Performing these calculations, as in relation 3.4, the third tableau is obtained.

TABLE 3.3 Third Tableau – Variable c Entered the Solution Basis Too, $Z = 1400$.

Basis	t	c	s_1	s_2	s_3	RHS
t	1	0	1	-0.5	0	20
c	0	1	-1	1	0	20
s_3	0	0	1	-1	1	10
Z	0	0	-10	-10	0	**-1400**

3.4.3 Third Tableau

Optimality test?

This tableau (Table 3.3) corresponds to the optimal solution! In fact, both objective function coefficients for the non-basic variables are negative, –10 and –10, and the objective function cannot be further improved upon.

Finally, as the non-basic variables take value 0, the basic variables take the respective RHS values, and the optimal solution follows:

$$\begin{cases} s_1 = 0 \\ s_2 = 0 \end{cases} \Rightarrow \begin{cases} t = 20 \\ c = 20 \\ s_3 = 10 \end{cases} \quad \text{then } Z = 1400$$

3.5 MATRIX FORM

In this section, the LP Simplex method is presented in matrix form, that is, by using matrix and vectors operations that correspond to the procedures developed within the algebraic and/or tableau form. At each iteration, obtaining the inverse matrix is an important step to solve the linear system of equations; note that the inverse matrix allows the calculation of multiple RHS and it paves the way for the analysis of sensitive parameters. In this way, the updating approach to obtain the inverse matrix is also important. The approach is presented and related to the corresponding algebraic procedures.

Now we address the LP Simplex through matrix and vectors operations. The LP in augmented form is

$$\max [Z] = \begin{bmatrix} \mathbf{c} & 0 \end{bmatrix} \cdot \begin{bmatrix} \mathbf{x} \\ \mathbf{x_s} \end{bmatrix}$$

subject to

$$\begin{bmatrix} \mathbf{A} & \mathbf{I} \end{bmatrix} \cdot \begin{bmatrix} \mathbf{x} \\ \mathbf{x_s} \end{bmatrix} = \mathbf{b} ; \quad \begin{bmatrix} \mathbf{x} \\ \mathbf{x_s} \end{bmatrix} \geq 0$$

For the furniture factory problem:

$$\max [Z] = \begin{bmatrix} 40 & 30 \end{bmatrix} \cdot \begin{bmatrix} t \\ c \end{bmatrix}$$

subject to

$$\begin{bmatrix} 2 & 1 \\ 2 & 2 \\ 0 & 1 \end{bmatrix} \cdot \begin{bmatrix} t \\ c \end{bmatrix} \le \begin{bmatrix} 60 \\ 80 \\ 30 \end{bmatrix}$$

$$\begin{bmatrix} t & c \end{bmatrix}^T \ge 0$$

The LP Simplex in matrix form presents the following steps:

1. Select the basic variables vector, $\mathbf{x_B}$.

2. Define the basic matrix, $\mathbf{B}$.

3. Invert the basic matrix, $\mathbf{B^{-1}}$.

4. Calculate the basic solution, $\mathbf{x_B} = \mathbf{B^{-1}.b}$; $Z = \mathbf{c_B.x_B}$.

5. Update the non-basic matrix: $\mathbf{K} = \mathbf{B^{-1}.A_N}$.

6. Update the objective function: $y = \mathbf{c_B.B^{-1}}$; $\mathbf{r} = \mathbf{c_N} - \mathbf{c_B.K}$.

7. Check optimality ($\mathbf{c_N} < 0$) or select the non-basic variable with the largest positive value (column e) as the entering variable.

8. Choose the variable associated with the minimum ratio value, $\theta_l = [(\mathbf{x_B})_l / k_{le}]$, as the leaving variable (line l); return to the first step.

When minimizing, consider the opposite situation in step 7: the solution is optimal if there are only positive coefficients; otherwise, select the negative coefficient with the largest absolute value to define the entering variable.

3.5.1 First Iteration

The LP problem in matrix form:

$$\max [Z] = \begin{bmatrix} 40 & 30 & | & 0 & 0 & 0 \end{bmatrix} \cdot \begin{bmatrix} t \\ c \\ s_1 \\ s_2 \\ s_3 \end{bmatrix}$$

subject to

$$\begin{bmatrix} 2 & 2 & | & 1 & 0 & 0 \\ 2 & 1 & | & 0 & 1 & 0 \\ 0 & 1 & | & 0 & 0 & 1 \end{bmatrix} \cdot \begin{bmatrix} t \\ c \\ s_1 \\ s_2 \\ s_3 \end{bmatrix} = \begin{bmatrix} 60 \\ 80 \\ 30 \end{bmatrix}$$

$$\begin{bmatrix} t & c & s_1 & s_2 & s_3 \end{bmatrix}^T \geq 0$$

1) Select the basic, X_B, and non-basic, X_N, variables.

The basic variables and the non-basic variables are, respectively,

$$X_B = \{s_1, s_2, s_3\}$$

$$X_N = \{t, c\}$$

The system's RHS corresponds to the column vector, **b**:

$$b = \begin{bmatrix} 60 \\ 80 \\ 30 \end{bmatrix}$$

2) Define the basic matrix, B.

$$B = \begin{bmatrix} 1 & 0 & 0 \\ 0 & 1 & 0 \\ 0 & 0 & 1 \end{bmatrix}$$

and also

$$\mathbf{A_N} = \begin{bmatrix} 2 & 1 \\ 2 & 2 \\ 0 & 1 \end{bmatrix}$$

In addition, the following are defined:

$$\mathbf{c_B} = \begin{bmatrix} 0 & 0 & 0 \end{bmatrix}$$

and

$$\mathbf{c_N} = \begin{bmatrix} 40 & 30 \end{bmatrix}$$

3) Invert the basic matrix, *Inv*(B).

For the first iteration, the inverse matrix is also the identity matrix:

$$\mathbf{B}^{-1} = Inv(\mathbf{B}) = \begin{bmatrix} 1 & 0 & 0 \\ 0 & 1 & 0 \\ 0 & 0 & 1 \end{bmatrix}$$

4) Calculate the basic solution.

$$\mathbf{x_b} = \mathbf{B}^{-1}.\mathbf{b} = \begin{bmatrix} 1 & 0 & 0 \\ 0 & 1 & 0 \\ 0 & 0 & 1 \end{bmatrix}.\begin{bmatrix} 60 \\ 80 \\ 30 \end{bmatrix} = \begin{bmatrix} 60 \\ 80 \\ 30 \end{bmatrix}$$

Then

$$Z = \mathbf{c_B}.\mathbf{x_B} = \begin{bmatrix} 0 & 0 & 0 \end{bmatrix}.\begin{bmatrix} 60 \\ 80 \\ 30 \end{bmatrix} = 0$$

5) Update the non-basic matrix.

$$K = B^{-1}.A_N = \begin{bmatrix} 1 & 0 & 0 \\ 0 & 1 & 0 \\ 0 & 0 & 1 \end{bmatrix} . \begin{bmatrix} 2 & 1 \\ 2 & 2 \\ 0 & 1 \end{bmatrix} == \begin{bmatrix} 2 & 1 \\ 2 & 2 \\ 0 & 1 \end{bmatrix}$$

6) Update the objective function.

Update the coefficients for the basic variables (corresponding to dual values, **y**):

$$y = c_B.B^{-1} = \begin{bmatrix} 0 & 0 & 0 \end{bmatrix} . \begin{bmatrix} 1 & 0 & 0 \\ 0 & 1 & 0 \\ 0 & 0 & 1 \end{bmatrix} = \begin{bmatrix} 0 & 0 & 0 \end{bmatrix}$$

Update the coefficients (reduced costs, **r**) for the non-basic variables:

$$r = c_N - c_B.K = \begin{bmatrix} 40 & 30 \end{bmatrix} - \begin{bmatrix} 0 & 0 & 0 \end{bmatrix} . \begin{bmatrix} 2 & 1 \\ 2 & 2 \\ 0 & 1 \end{bmatrix}$$

$$= \begin{bmatrix} 40 & 30 \end{bmatrix} - \begin{bmatrix} 0 & 0 \end{bmatrix}$$

$$= \begin{bmatrix} 40 & 30 \end{bmatrix}$$

7) Optimality?

It can be observed that the current solution is not optimal because the reduced costs are positive and the objective function can be improved.

o The first variable, t, enters the basic set (let it be, $e = 1$), since it presents the highest positive value, 40, for the non-basic variables in the objective function.

8) Apply the minimum ratio test.

As usual, the ratio between the resources parameters on the RHS and the coefficients for the entering variable in the first column ($e = 1$) is

$$\begin{cases} \theta_1 = \dfrac{60}{2} = 30 \\[2mm] \theta_2 = \dfrac{80}{2} = 40 \\[2mm] \theta_3 = \dfrac{30}{0} \ n.d. \end{cases}$$

o The limiting resource is presented in the first restriction (let it be, $l = 1$) and the slack variable, s_1, leaves the basic variables set.

Note again, the next solution presents a value improved by

$$\Delta Z = c_1 \cdot \theta_1$$

$$= 40 \times 30 = 1200$$

3.5.2 Second Iteration

1) **Select the basic, X_B, and non-basic, X_N, variables.**

Now, swap the variables t and s_1; that is, t enters the basic variables set while s_1 leaves it and moves onto the non-basic variables set:

$$X_B = \{t, s_2, s_3\}$$

$$X_N = \{s_1, c\}$$

2) **Define the basic matrix, B.**

By swapping variables t and s_1, the associated columns in the basic matrix ($l = 1$) and the non-basic matrix ($e = 1$) also swap; note that the column ordering is retained in both matrix:

$$B = \begin{bmatrix} 2 & 0 & 0 \\ 2 & 1 & 0 \\ 1 & 0 & 1 \end{bmatrix}$$

and also

$$A_N = \begin{bmatrix} 1 & 1 \\ 0 & 2 \\ 0 & 1 \end{bmatrix}$$

In addition, the coefficient vectors are also updated:

$$c_B = \begin{bmatrix} 40 & 0 & 0 \end{bmatrix}$$

and

$$c_N = \begin{bmatrix} 0 & 30 \end{bmatrix}$$

3) Invert the basic matrix, *Inv*(B).

The inverse matrix can be obtained by automatic calculation, either using intrinsic functions on calculators or computer applications, or manually with an updating procedure (presented later in this section).

$$B^{-1} = Inv(B) = \begin{bmatrix} 0.5 & 0 & 0 \\ -1 & 1 & 0 \\ 0 & 0 & 1 \end{bmatrix}$$

4) Calculate the basic solution.

$$x_b = B^{-1} \cdot b = \begin{bmatrix} 0.5 & 0 & 0 \\ -1 & 1 & 0 \\ 0 & 0 & 1 \end{bmatrix} \cdot \begin{bmatrix} 60 \\ 80 \\ 30 \end{bmatrix}$$

$$= \begin{bmatrix} 0.5 \times 60 + 0 \times 80 + 0 \times 30 \\ -1 \times 60 + 1 \times 80 + 0 \times 30 \\ 0 \times 60 + 0 \times 80 + 1 \times 30 \end{bmatrix} = \begin{bmatrix} 30 + 0 + 0 \\ -60 + 80 + 0 \\ 0 + 0 + 30 \end{bmatrix} = \begin{bmatrix} 30 \\ 20 \\ 30 \end{bmatrix}$$

Then

$$Z = c_B \cdot x_B = \begin{bmatrix} 40 & 0 & 0 \end{bmatrix} \cdot \begin{bmatrix} 30 \\ 20 \\ 30 \end{bmatrix} = 1200$$

5) **Update the non-basic matrix.**

$$K = B^{-1} \cdot A_N = \begin{bmatrix} 0.5 & 0 & 0 \\ -1 & 1 & 0 \\ 0 & 0 & 1 \end{bmatrix} \cdot \begin{bmatrix} 1 & 1 \\ 0 & 2 \\ 0 & 1 \end{bmatrix}$$

$$= \begin{bmatrix} 0.5 \times 1 + 0 \times 0 + 0 \times 0 & 0.5 \times 1 + 0 \times 2 + 0 \times 1 \\ -1 \times 1 + 1 \times 0 + 0 \times 0 & -1 \times 1 + 1 \times 2 + 0 \times 1 \\ 0 \times 1 + 0 \times 0 + 1 \times 0 & 0 \times 1 + 0 \times 2 + 1 \times 1 \end{bmatrix} = \begin{bmatrix} 0.5 & 0.5 \\ -1 & 1 \\ 0 & 1 \end{bmatrix}$$

6) **Update the objective function.**

Update the coefficients for basic variables:

$$y = c_B \cdot B^{-1} = \begin{bmatrix} 40 & 0 & 0 \end{bmatrix} \cdot \begin{bmatrix} 0.5 & 0 & 0 \\ -1 & 1 & 0 \\ 0 & 0 & 1 \end{bmatrix} = \begin{bmatrix} 20 & 0 & 0 \end{bmatrix}$$

Update the coefficients for non-basic variables:

$$r = c_N - c_B \cdot K = \begin{bmatrix} 0 & 30 \end{bmatrix} - \begin{bmatrix} 40 & 0 & 0 \end{bmatrix} \cdot \begin{bmatrix} 0.5 & 0.5 \\ -1 & 1 \\ 0 & 1 \end{bmatrix}$$

$$= \begin{bmatrix} 0 & 30 \end{bmatrix} - \begin{bmatrix} 20 & 20 \end{bmatrix}$$

$$= \begin{bmatrix} -20 & 10 \end{bmatrix}$$

7) **Optimality?**

The current solution is still not optimal because the reduced cost for variable c remains positive, 10, thus the objective function can be improved; for that, the second non-basic variable enters the basic set (let it be $e = 2$).

8) **Apply the minimum ratio test.**

As usual, the ratio between the resources parameters on the RHS and the coefficients for the entering variable in the associated column is

$$
\left\{
\begin{array}{l}
\theta_1 = \dfrac{30}{0.5} = 60 \\[2mm]
\theta_2 = \dfrac{20}{1} = 20 \\[2mm]
\theta_3 = \dfrac{30}{1} = 30
\end{array}
\right.
$$

- The limiting resource is now in the second restriction (let it be $l = 2$) and the slack variable, s_2, leaves the basic variables set.

Addressing the new basic variable, c, the next solution will be improved by

$$\Delta Z = c_2 \cdot \theta_2$$

$$= 10 \times 20 = 200$$

3.5.3 Third Iteration

1) **Select the basic, X_B, and non-basic, X_N, variables.**

In this iteration, swap variables c and s_2; that is, c enters the basic variables set while s_2 moves to the non-basic variables set:

$$X_B = \{t, c, s_3\}$$

$$X_N = \{s_1, s_2\}$$

2) **Define the basic matrix, B.**

By swapping variables c and s_2, the associated columns in the basic matrix ($l = 2$) and the non-basic matrix ($e = 2$) also swap; again, the column ordering is retained in both matrix:

$$
B = \begin{bmatrix} 2 & 1 & 0 \\ 2 & 2 & 0 \\ 0 & 1 & 1 \end{bmatrix}
$$

and also

$$A_N = \begin{bmatrix} 1 & 0 \\ 0 & 1 \\ 0 & 0 \end{bmatrix}$$

In addition, the coefficient vectors are also updated:

$$c_B = \begin{bmatrix} 40 & 30 & 0 \end{bmatrix}$$

and

$$c_N = \begin{bmatrix} 0 & 0 \end{bmatrix}$$

3) Invert the basic matrix, *Inv*(B).

The inverse matrix can be obtained by automatic calculation or manually using an updating procedure.

$$B^{-1} = Inv(B) = \begin{bmatrix} 1 & -0.5 & 0 \\ -1 & 1 & 0 \\ 1 & -1 & 1 \end{bmatrix}$$

4) Calculate the basic solution.

$$x_b = B^{-1}.b = \begin{bmatrix} 1 & -0.5 & 0 \\ -1 & 1 & 0 \\ 1 & -1 & 1 \end{bmatrix} . \begin{bmatrix} 60 \\ 80 \\ 30 \end{bmatrix}$$

$$= \begin{bmatrix} 1 \times 60 - 0.5 \times 80 + 0 \times 30 \\ -1 \times 60 + 1 \times 80 + 0 \times 30 \\ 1 \times 60 - 1 \times 80 + 1 \times 30 \end{bmatrix} = \begin{bmatrix} 60 - 40 + 0 \\ -60 + 80 + 0 \\ 60 - 80 + 30 \end{bmatrix} = \begin{bmatrix} 20 \\ 20 \\ 10 \end{bmatrix}$$

Then

$$Z = c_B \cdot x_B = \begin{bmatrix} 40 & 30 & 0 \end{bmatrix} \cdot \begin{bmatrix} 20 \\ 20 \\ 10 \end{bmatrix} = 800 + 600 + 0 = 1400$$

5) Update the non-basic matrix.

$$\mathbf{K} = \mathbf{B}^{-1} \cdot \mathbf{A_N} = \begin{bmatrix} 1 & -0.5 & 0 \\ -1 & 1 & 0 \\ 1 & -1 & 1 \end{bmatrix} \cdot \begin{bmatrix} 1 & 0 \\ 0 & 1 \\ 0 & 0 \end{bmatrix}$$

$$= \begin{bmatrix} 1\times1 - 0.5\times0 + 0\times0 & 1\times0 - 0.5\times1 + 0\times0 \\ -1\times1 + 1\times0 + 0\times0 & -1\times0 + 1\times1 + 0\times0 \\ 1\times1 - 1\times0 + 1\times0 & 1\times0 - 1\times1 + 1\times0 \end{bmatrix} = \begin{bmatrix} 1 & -0.5 \\ -1 & 1 \\ 1 & -1 \end{bmatrix}$$

6) Update the objective function.

Update the coefficients for the basic variables:

$$\mathbf{y} = c_B \cdot \mathbf{B}^{-1} = \begin{bmatrix} 40 & 30 & 0 \end{bmatrix} \cdot \begin{bmatrix} 1 & -0.5 & 0 \\ -1 & 1 & 0 \\ 1 & -1 & 1 \end{bmatrix} = \begin{bmatrix} 10 & 10 & 0 \end{bmatrix}$$

Update the coefficients for the non-basic variables:

$$\mathbf{r} = c_N - c_B \cdot \mathbf{K} = \begin{bmatrix} 0 & 0 \end{bmatrix} - \begin{bmatrix} 40 & 30 & 0 \end{bmatrix} \cdot \begin{bmatrix} 1 & -0.5 \\ -1 & 1 \\ 1 & -1 \end{bmatrix}$$

$$= \begin{bmatrix} 0 & 0 \end{bmatrix} - \begin{bmatrix} 10 & 10 \end{bmatrix}$$

$$= \begin{bmatrix} -10 & -10 \end{bmatrix}$$

7) **Optimality?**

The current solution is optimal!

o The objective function cannot be improved because the reduced costs for the non-basic variables are both negative.

3.6 UPDATING THE INVERSE MATRIX

The inversion of the basic matrix, $Inv(\mathbf{B})$, is efficiently computed by updating the inverse matrix from the last iteration:

$$\left(\mathbf{B}^{-1}\right)_{(i+1)} = \mathbf{U}.\left(\mathbf{B}^{-1}\right)_{(i)}$$

The updating matrix $\mathbf{U}$ is defined as the identity matrix with column e modified:

$$\mathbf{U} = \begin{bmatrix} 1 & 0 & 0 & ... & -k_{1l}/k_{le} & ... & 0 \\ 0 & 1 & 0 & ... & -k_{2l}/k_{le} & ... & 0 \\ 0 & 0 & 1 & ... & -k_{3l}/k_{le} & ... & 0 \\ & & & (...) & & & \\ 0 & 0 & 0 & ... & 1/k_{le} & ... & 0 \\ & & & (...) & & & \\ 0 & 0 & 0 & ... & -k_{ml}/k_{le} & ... & 1 \end{bmatrix}$$

Following on from the matrix Simplex procedure in the previous section, indicator e is related to the ordering of the entering variable in each iteration, while indicator l is related to the leaving variable.

In comparing with the algebraic (or tableau) form, note that

- The pivot in each iteration is defined by the column of the entering variable, e, and the line of the leaving variable, l, corresponding to the coefficient k_{le}.

- Then, all the coefficients at line l are divided by the pivot coefficient, or better, they are multiplied by its inverse $(1/k_{le})$.

- Multiplier factors are used for all other lines $(1, 2, 3,..., m)$ in such a way as to properly subtract the pivot line $(-k_{11}/k_{le}, -k_{21}/k_{le}, -k_{31}/k_{le},..., -k_{ml}/k_{le})$ and proceed with Gauss elimination.

From the first iteration in matrix form, the entering variable's column, $e = 1$, and the leaving variable's line, $l = 1$, are defined; then,

- The pivot coefficient used in the second iteration is $k_{11} = 2$.

- All the coefficients on line one are divided by the pivot coefficient, that is, they are multiplied by $(1/2)$.

- Multiplier factors are used for the other lines (two and three), in such a way as to properly subtract the pivot line $(-2/2, -0/2)$ and proceed with Gauss elimination.

- The updated coefficients can be found in matrix $\mathbf{K}$, at step 5; however, in the first iteration they are trivially equal to the $\mathbf{A_N}$ coefficients.

Then

$$U = \begin{bmatrix} \frac{1}{2} & 0 & 0 \\ -\frac{2}{2} & 1 & 0 \\ -\frac{0}{2} & 0 & 1 \end{bmatrix}$$

And the inverse matrix for the second step can be obtained by updating the inverse matrix in the first step:

$$\left(B^{-1}\right)_{(2)} = U \cdot \left(B^{-1}\right)_{(1)}$$

Since the inverse matrix in the first step is the identity matrix, $\mathbf{I}$, then the updating matrix also becomes the associated result:

$$\left(\mathbf{B}^{-1}\right)_{(2)} = \begin{bmatrix} \frac{1}{2} & 0 & 0 \\ -\frac{2}{2} & 1 & 0 \\ -\frac{0}{2} & 0 & 1 \end{bmatrix} \cdot \begin{bmatrix} 1 & 0 & 0 \\ 0 & 1 & 0 \\ 0 & 0 & 1 \end{bmatrix} = \begin{bmatrix} 0.5 & 0 & 0 \\ -1 & 1 & 0 \\ 0 & 0 & 1 \end{bmatrix}$$

The inverse matrix can be observed in the columns associated with the slack variables, either in algebraic or in tableau form; namely, the transformation to the second basic solution can be found in relation 3.3 and in the second tableau (Table 3.2), respectively.

- At this point, the updating matrix $\mathbf{U}$ and the inverse matrix are equal; for that, they both show that during Gauss elimination, line one was divided by 2, it was directly subtracted from line two, and it did not affect line three, as shown in the matrix's first column.

And from the second iteration in matrix form, the entering variable's column, $e = 2$, and the leaving variable's line, $l = 2$, are defined; then,

- The pivot coefficient used for the third iteration is $k_{22} = 1$.

- All the coefficients on line two are divided by the pivot coefficient, that is, they are multiplied by $(1/1)$.

- Multiplier factors are used for the other lines (one and three), in such a way as to properly subtract the pivot line $(-0.5/1, -1/1)$.

- Note that the updated coefficients can be found by reading the second column of matrix $\mathbf{K}$, in the fifth step of the second iteration:

$$\mathbf{U} = \begin{bmatrix} 1 & -0.5/1 & 0 \\ 0 & 1/1 & 0 \\ 0 & -1/1 & 1 \end{bmatrix}$$

By updating the inverse matrix in the second step, the inverse matrix for the third step is thus calculated:

$$\left(\mathbf{B}^{-1}\right)_{(3)} = \mathbf{U} \cdot \left(\mathbf{B}^{-1}\right)_{(2)}$$

Finally,

$$\left(\mathbf{B}^{-1}\right)_{(3)} = \begin{bmatrix} 1 & -0.5 & 0 \\ 0 & 1 & 0 \\ 0 & -1 & 1 \end{bmatrix} \cdot \begin{bmatrix} 0.5 & 0 & 0 \\ -1 & 1 & 0 \\ 0 & 0 & 1 \end{bmatrix} = \begin{bmatrix} 1 & -0.5 & 0 \\ -1 & 1 & 0 \\ 1 & -1 & 1 \end{bmatrix}$$

Again, this inverse matrix can be observed in the columns associated with the slack variables, either in algebraic or in tableau form; namely, it can be found in relation 3.4 and in the third tableau (Table 3.3), respectively.

- The updating matrix $\mathbf{U}$ and the inverse matrix are not equal, but they are coincident in the second column; they both show that in the second Gauss elimination, line two was the pivot line and remained the same (divided by 1), it was directly subtracted from line three, and it was divided by 2 and subtracted from line one.

3.7 CONCLUDING REMARKS

In this chapter, the LP basics are revisited, including the graphical approach. The LP Simplex method is presented in different but closely connected forms: the algebraic, the tabular, and the matrix form. Updating the inverse matrix is also focused on, since the inverse matrix is strongly connected with the referred LP forms and it also supports important developments.

LP-based modeling drives better solutions than LA-based modeling; slack variables are introduced and the LP improvements are both qualitative and quantitative:

- Degrees of freedom for better decision-making are allowed through the introduction of slack variables.

- Slack variables provide an enlarged search space, with one additional dimension for each variable.

- Also, only non-negative solutions are allowed in LP; that is, negative solutions are not allowed, thus real-world applications requiring non-negative values are promoted.

- LA approaches and methods are integrated at each iteration of the LP Simplex method, either in algebraic or in matrix form.

- The matrix form and the inverse matrix for LP basic variables support the treatment of multiple RHS; thereafter, analyzing the alterations of sensitive RHS parameters is direct.

- Sensitivity analysis and uncertainty factors are treated in Chapter 6; namely, by successively addressing the alterations either in the objective function coefficients or in the RHS parameters, or by introducing new variables.

The LP framework for decision-making is thus enlarged and encapsulates the LA approaches. However, implicitly, LP is assuming that one decision maker selects the optimal solution in concordance with the objective function value, that is, one alternative is selected by considering one single attribute.

Duality and related concepts are important for LP-based decision-making, namely, by assuming a second decision maker. Important topics, such as primal-dual transformation, primal-dual relations, and dual economic interpretation are addressed in Chapter 4. In addition, duality concepts are also important for LP optimality analysis, Game Theory, and other advanced features that are relevant in robust decision-making.

Duality

THIS CHAPTER INTRODUCES DUALITY and the complementary dual problem, including a simple transformation from the primal linear programming (LP) problem, the Simplex tableau to obtain the optimal dual solution, as well as the main properties and the economic interpretation of dual results. Noting that the primal LP problem is complemented and enlarged in the non-feasible region through the associated dual problem, a new insight into optimality analysis and sensitive parameters is obtained, and decision-making's attributes are improved.

4.1 INTRODUCTION

In close relation with each LP problem, the primal problem, there is another problem called the dual LP. The dual LP can be interpreted as a complementary form of the primal LP, and vice versa, with important attributes for decision support. In fact, the dual variables correspond to additional information about resource utilization; additionally, the complementary properties for the primal and dual solutions are very important, either in optimal value or in suboptimal points.

The dual LP problem is obtained from the corresponding primal LP, namely:

- The primal LP aims at the n-variables objective function maximization with m restrictions of the "less than or equal to" type; complementary, the dual LP aims at the m-variables objective minimization with n restrictions of the "greater than or equal to" type.

DOI: 10.1201/9781315200323-4

- Each primal restriction corresponds to a dual variable, in a one-to-one relation; and vice versa, each variable in the primal LP, respectively, corresponds to a restriction in the dual problem.

- The objective function coefficients for the m dual variables are obtained from the m right-hand side (RHS) parameters in the primal restrictions; and vice versa, the n dual RHS parameters are obtained from the n primal coefficients in the objective function.

- By matrix transposition, the restrictions matrix ($n \times m$) for the dual LP is obtained from the restrictions matrix ($m \times n$) for the primal LP (and vice versa).

Primal	Dual

$$\begin{bmatrix} \max [Z] = \mathbf{c.x} \\ \text{subject to} \\ \mathbf{A.x} \le \mathbf{b} \\ \mathbf{x} \ge 0 \end{bmatrix} \begin{array}{c} \rightarrow \\ \leftarrow \end{array} \begin{bmatrix} \min [W] = \mathbf{y.b} \\ \text{subject to} \\ \mathbf{A}^{\mathsf{T}}.\mathbf{y} \ge \mathbf{c} \\ \mathbf{y} \ge 0 \end{bmatrix}$$

4.2 PRIMAL-DUAL TRANSFORMATIONS

In general, it is assumed that the dual transformation of a dual LP results in the primal LP again. It is also assumed that each primal restriction is associated with a dual variable, and that each primal variable is associated with a dual restriction.

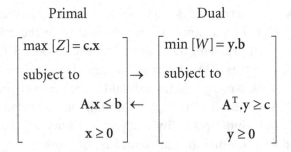

$$\left\{ \begin{array}{l} \max \\ \text{restriction } (i) \le \\ \text{restriction } (j) \ge \\ \text{restriction } (k) = \\ x_l \ge 0 \\ x_m \le 0 \\ \text{non-constrained } x_n \end{array} \right\} \begin{array}{c} \rightarrow \\ \leftarrow \end{array} \left\{ \begin{array}{l} \min \\ y_i \ge 0 \\ y_j \le 0 \\ \text{non-constrained } y_k \\ \text{restriction } (l) \ge \\ \text{restriction } (m) \le \\ \text{restriction } (n) = \end{array} \right\}$$

For example, let it be the primal problem:

$$\max Z = 40t + 30c$$

subject to

$$\begin{cases} 2t + 1c \leq 60 \\ 2t + 2c \leq 80 \\ 1c \leq 30 \end{cases}$$

A suggestion for the primal-dual transformation is to consider using the LP matrix form for the primal problem, as described in Chapter 3:

$$\max [Z] = \begin{bmatrix} 40 & 30 \end{bmatrix} \cdot \begin{bmatrix} t \\ c \end{bmatrix}$$

subject to

$$\begin{bmatrix} 2 & 1 \\ 2 & 2 \\ 0 & 1 \end{bmatrix} \cdot \begin{bmatrix} t \\ c \end{bmatrix} \leq \begin{bmatrix} 60 \\ 80 \\ 30 \end{bmatrix}$$

$$\begin{bmatrix} t & c \end{bmatrix}^T \geq 0$$

Synoptically, the transformation primal-dual for this LP problem considers the following steps:

- If the primal LP is a maximization problem with "less than or equal to" ($\leq$) restrictions, then the associated dual is a minimization problem with "greater than or equal to" ($\geq$) restrictions.

- The two variables $\{t, c\}$ in the primal LP are being associated with two restrictions in the dual problem, in a one-to-one relation; and vice versa, the three primal restrictions, respectively, are also associated with the three dual variables $\{y_1, y_2, y_3\}$.

- The coefficients for the primal variables in the objective function (40, 30) move to the RHS of the associated dual restrictions; and vice versa, the primal RHS parameters (60, 80, 30) move to the dual objective function coefficients; in matrix form, this can easily be done by swapping and transposing the associated vectors for the objective function's coefficients and RHS parameters.

- By transposition too, the restrictions matrix for the primal problem originates the restrictions matrix for the dual problem.

The matrix form obtained in this way for the dual problem is

$$\min \ [W] = \begin{bmatrix} 60 & 80 & 30 \end{bmatrix} \cdot \begin{bmatrix} y_1 \\ y_2 \\ y_3 \end{bmatrix}$$

subject to

$$\begin{bmatrix} 2 & 2 & 0 \\ 1 & 2 & 1 \end{bmatrix} \cdot \begin{bmatrix} y_1 \\ y_2 \\ y_3 \end{bmatrix} \geq \begin{bmatrix} 40 \\ 30 \end{bmatrix}$$

$$\begin{bmatrix} y_1 & y_2 & y_3 \end{bmatrix}^T \geq 0$$

Then, the dual LP problem in algebraic form follows directly:

$$\min W = 60y_1 + 80y_2 + 30y_3$$

subject to

$$\begin{cases} 2y_1 + 2y_2 & \geq 40 \\ 1y_1 + 2y_2 + 1y_3 \geq 30 \end{cases}$$

$$y_1, y_2, y_3 \geq 0$$

4.3 DUAL SIMPLEX METHOD

In manipulating the dual minimization LP problem:

$$\min W = 60y_1 + 80y_2 + 30y_3$$

subject to

$$\begin{cases} 2y_1 + 2y_2 & \geq 40 \\ 1y_1 + 2y_2 + 1y_3 \geq 30 \end{cases}$$

$$y_1, y_2, y_3 \geq 0$$

by introducing the restrictions' dual variables, the excess variables r_1 and r_2, a linear system of equations with five variables and two restrictions is obtained:

$$\min W = 60y_1 + 80y_2 + 30y_3 + 0r_1 + 0r_2$$

subject to

$$\begin{cases} 2y_1 + 2y_2 & -1r_1 & = 40 \\ 1y_1 + 2y_2 + 1y_3 & -1r_2 = 30 \end{cases}$$

$$y_1, y_2, y_3, r_1, r_2 \geq 0$$

The dual minimization problem can be transformed into a maximization problem by multiplying all relations by –1:

$$\max(-W) = -60y_1 - 80y_2 - 30y_3 + 0r_1 + 0r_2$$

subject to

$$\begin{cases} -2y_1 - 2y_2 & +1r_1 & = -40 \\ -1y_1 - 2y_2 - 1y_3 & +1r_2 = -30 \end{cases}$$

$$y_1, y_2, y_3, r_1, r_2 \geq 0$$

To solve this linear system, three variables are made zero (the non-basic variables), and then a system with two (basic) variables can be solved using two equations.

4.3.1 First Iteration

Assuming:

$$\begin{cases} y_1 = 0 \\ y_2 = 0 \\ y_3 = 0 \end{cases} \Rightarrow \begin{cases} r_1 = -40 \\ r_2 = -30 \end{cases} \quad \text{then } (-W) = 0$$

This solution is not possible since it presents negative values for variables r_1 and r_2; the most violated restriction is in the first line, with a negative value of -40, and the pivot column associated is defined by the minimum ratio test.

For the non-basic variables $\{y_1, y_2, y_3\}$, the minimum ratio is calculated through the division of the corresponding coefficients in the objective function and in the pivot line:

$$\begin{cases} \theta_1 = {}^{-60}\!/_{-2} = 30 \\ \theta_2 = {}^{-80}\!/_{-2} = 40 \\ \theta_3 = {}^{-30}\!/_{0} \quad n.d. \end{cases}$$

Then, the variable y_1 enters the basis while the dual variable r_1 exits the basis, and the pivot element (-2) is obtained in line one, column one. In this way, the next solution will present a value improved by

$$\Delta W = r_1 \cdot \theta_1$$

$$= -40 \times 30 = -1200$$

Addressing the new basic variable, y_1, by Gauss elimination:

- Dividing line one by -2.
- Multiplying line one by $-1/2$ (i.e., dividing by -2), and adding to line two.
- Multiplying line one by 30 and subtracting from the objective line.

The next linear system is

$$\max(-W) = 0\,y_1 - 20\,y_2 - 30\,y_3 - 30\,r_1 + 0\,r_2$$

subject to

$$\begin{cases} 1\,y_1 + 1\,y_2 & -0.5\,r_1 & = 20 \\ 0\,y_1 - 1\,y_2 - 1\,y_3 - 0.5\,r_1 + 1\,r_2 & = -10 \end{cases}$$

$$y_1, y_2, y_3, r_1, r_2 \geq 0$$

4.3.2 Second Iteration

Assuming:

$$\begin{cases} r_1 = 0 \\ y_2 = 0 \\ y_3 = 0 \end{cases} \Rightarrow \begin{cases} y_1 = 20 \\ r_2 = -10 \end{cases} \quad \text{then } (-W) = -1200$$

This solution is not possible because it still presents a negative value for variable r_2, −10, in the second line; again, the pivot column is to be defined by the minimum ratio test.

For the non-basic variables $\{y_2, y_3, r_1\}$, respectively, the division of the correspondent coefficients in the objective function and in the second line is

$$\begin{vmatrix} \theta_1 = {-20}/{-1} = 20 \\ \theta_2 = {-30}/{-1} = 30 \\ \theta_3 = \dfrac{-30}{(-1/2)} = 60 \end{vmatrix}$$

Then, dual variable y_2 enters the basis while variable r_2 exits the basis, and the pivot element (−1) is obtained in line two, column two. The next solution will present a value improved by

$$\Delta W = r_2 \cdot \theta_1$$

$$= -10 \times 20 = -200$$

Addressing the new basic variable, y_2, by Gauss elimination:

- Dividing (or multiplying, in this case) line two by –1.
- Directly adding line two to line one.
- Multiplying line two by 20 and subtracting from the objective line.

The next linear system represents the optimal solution:

$$\max(-W) = 0\, y_1 + 0\, y_2 - 10 y_3 - 20 r_1 - 20 r_2$$

subject to

$$\begin{cases} 1 y_1 + 0 y_2 - 1 y_3 - 1 r_1 + 1 r_2 = 10 \\ 0 y_1 + 1 y_2 + 1 y_3 + 0.5\, r_1 - 0.5 r_2 = 10 \end{cases}$$

$$y_1, y_2, y_3, r_1, r_2 \geq 0$$

This solution still satisfies the optimality criteria, due to the negative coefficients on the objective function for the non-basic variables. However, this solution is already in the feasible region, the dual basic variables are now positive.

The dual solution is

$$\begin{cases} r_1 = 0 \\ r_2 = 0 \\ y_3 = 0 \end{cases} \Rightarrow \begin{cases} y_1 = 10 \\ y_2 = 10 \end{cases} \quad \text{then} \quad (-W) = -1400$$

This is the optimal solution for the dual problem!

4.3.3 Dual Simplex in Tableau Form

The corresponding tableau for the dual problem solved in algebraic form follow.

- **First tableau**

In the first tableau, typically, the dual decision variables are assumed 0 (non-basic variables), while the excess variables are selected as basic variables, and the objective function's value is $W = 0$ (Table 4.1).

TABLE 4.1   First Tableau – Initial Basic Solution, $W = 0$.

Basis	y_1	y_2	y_3	r_1	r_2	RHS
r_1	-2	-2	0	1	0	-40
r_2	-1	-2	-1	0	1	-30
$-W$	-60	-80	-30	0	0	0

First restriction is the more violated restriction, then r_1 exits the solution basis because it presents the most negative value, –40.

The non-basic variable, y_1, enters the solution basis in the sequence of the minimum ratio obtained for the non-basic variables $\{y_1, y_2, y_3\}$, with $y_1 = 30$:

$$\begin{cases} \theta_1 = {}^{-60}\!/_{-2} = 30 \\ \theta_2 = {}^{-80}\!/_{-2} = 40 \\ \theta_3 = {}^{-30}\!/_{0} \quad n.d. \end{cases}$$

In this way, the pivot element is –2 (gray background in first line, first column), and the next solution will be improved by 1200 (40 × 30) with the usual Gauss elimination procedures:

o Dividing line one by –2.

o Multiplying line one by –1/2 and adding to line two.

o Multiplying line one by 30 and subtracting from the objective line.

• **Second tableau**

The second tableau presents the internal cost for the first resource $y_1 = 20$, corresponding to the full utilization of 60 big-blue pieces in a total valorization of resources $W = 1200$ (Table 4.2).

TABLE 4.2 Second Tableau – Variable y_1 Entered the Solution Basis, $W = 1200$.

Basis	y_1	y_2	y_3	r_1	r_2	RHS
r_1	1	1	0	−0.5	0	20
r_2	0	−1	−1	−0.5	1	−10
−W	0	−20	−30	−30	0	−1200

However, the second restriction is still violated, r_2 presents a negative value, −10, thus it exits the basis.

Similarly, y_2 enters the basis when applying the minimum ratio test to the non-basic variables $\{y_2, y_3, r_1\}$, obtaining $y_2 = 20$:

$$\begin{cases} \theta_1 = \dfrac{-20}{-1} = 20 \\[2mm] \theta_2 = \dfrac{-30}{-1} = 30 \\[2mm] \theta_3 = \dfrac{-30}{(-1/2)} = 60 \end{cases}$$

Thus, the pivot element is −1 (gray background in second line, second column), and the next solution will be improved by 200 (10×20) within the usual Gauss elimination:

o Multiplying line two by −1.

o Directly adding line two to line one.

o Multiplying line two by 20 and subtracting from the objective line.

• **Third tableau**

The third tableau presents the internal costs both for the first resource, $y_1 = 10$, corresponding to the full utilization of 60 big-blue pieces, and for the second resource, $y_2 = 10$, also corresponding to the full utilization of 80 small-red pieces; the total valorization for resource utilization is thus $W = 1400$ (Table 4.3).

The current solution is feasible, and it is the optimal solution too, since no additional improvements to the objective function can be made!

TABLE 4.3   Third Tableau – Variable y_2 Entered the Solution Basis Too, $W = 1400$.

Basis	y_1	y_2	y_3	r_1	r_2	RHS
y_1	1	0	-1	-1	1	10
y_2	0	1	1	0.5	-0.5	10
$-W$	0	0	-10	-20	-20	**-1400**

Note that the optimal solution for the primal problem can be obtained from the objective function's line, by considering the symmetric values for the corresponding coefficients, i.e., for the associated variables. Namely:

- The decision variables $\{t, c\}$ for the primal problem are associated with the excess variables in the dual problem $\{r_1, r_2\}$, then $(t, c)^* = (20, 20)$.

- The primal slack variables $\{s_1, s_2, s_3\}$ are associated with the dual decision variables $\{y_1, y_2, y_3\}$, then $(s_1, s_2, s_3)^* = (0, 0, 10)$.

In this way, the complementary dual for the dual problem will result in the primal problem. In fact, the optimal solution for the dual problem is

$$\begin{bmatrix} y_1 \\ y_2 \\ y_3 \\ r_1 \\ r_2 \end{bmatrix}^* = \begin{bmatrix} 10 \\ 10 \\ 0 \\ 0 \\ 0 \end{bmatrix}$$

and for the primal problem:

$$\begin{bmatrix} s_1 \\ s_2 \\ s_3 \\ t \\ c \end{bmatrix}^* = \begin{bmatrix} 0 \\ 0 \\ 10 \\ 20 \\ 20 \end{bmatrix}$$

Thus, when the dual variables present positive values, the associated primal variables take 0; and when the dual variables take 0, the associated primal variables are positive. And vice versa, it is observed that when the primal variables are positive, the associated dual variables take 0; and when the primal variables take 0, the associated dual variables are positive. This set of observations is related to the duality properties for complementary primal-dual solutions, which are detailed in the next section.

4.4 DUALITY PROPERTIES

The optimal solution for the dual problem can also be found in the final tableau for the primal problem in Chapter 3 (algebraic form and tableau form in Sections 3.3 and 3.4, respectively; and matrix form in Section 3.5). It can be observed that

$\mathbf{y}^* = [10\ 10\ 0]$, internal costs, shadow prices.

$\mathbf{r}^* = [0\ 0]$, reduced costs.

The duality properties can be directly verified in this sub-section. Similarly, any intermediate solutions for both the primal and dual problems can be compared and verified.

- **Strong duality:**

    For the optimal solution, the objective value for the primal problem is equal to the objective value for the dual problem.

    $$Z^* = W^*$$

    $$\mathbf{c}.\mathbf{x}^* = \mathbf{y}^*.\mathbf{b}$$

    In fact, for the primal problem it can be verified that

    $$\begin{bmatrix} c_t & c_c & c_{s_3} \end{bmatrix}_\mathbf{B} \cdot \mathbf{x_B}^* = \begin{bmatrix} 40 & 30 & 0 \end{bmatrix} \cdot \begin{bmatrix} 20 \\ 20 \\ 10 \end{bmatrix}^* = 1400$$

    and also for the dual problem:

$$\begin{bmatrix} y_1 & y_2 & y_3 \end{bmatrix}^* . b = \begin{bmatrix} 10 & 10 & 0 \end{bmatrix}^* . \begin{bmatrix} 60 \\ 80 \\ 30 \end{bmatrix} = 1400$$

- **Complementary optimal solutions:**

At the final iteration of the Simplex method, optimal and complementing solutions are found.

In fact, the optimal solution for the primal problem is

$$\begin{bmatrix} t \\ c \\ s_1 \\ s_2 \\ s_3 \end{bmatrix}^* = \begin{bmatrix} 20 \\ 20 \\ 0 \\ 0 \\ 10 \end{bmatrix}$$

and for the dual problem:

$$\begin{bmatrix} r_1 \\ r_2 \\ y_1 \\ y_2 \\ y_3 \end{bmatrix}^* = \begin{bmatrix} 0 \\ 0 \\ 10 \\ 10 \\ 0 \end{bmatrix}$$

Comparing the two optimal solutions:

- o When the decision variables $\{t, c\}$ for the primal problem are positive, then the complementary dual variables $\{r_1, r_2\}$ are 0.

- o Similarly, when a slack variable for the primal problem is positive, s_3, then the associated dual variable, y_3, is 0; and when the slack variables for the primal problem $\{s_1, s_2\}$ are 0, the associated dual values $\{y_1, y_2\}$ are positive.

- **Complementary of feasible solutions:**

  At each iteration of the Simplex algorithm, feasible and complementing solutions are found.

  In the Simplex method, the first solution for the primal problem is

  $$\begin{bmatrix} t \\ c \\ s_1 \\ s_2 \\ s_3 \end{bmatrix} = \begin{bmatrix} 0 \\ 0 \\ 60 \\ 80 \\ 30 \end{bmatrix}$$

  and for the dual problem:

  $$\begin{bmatrix} r_1 \\ r_2 \\ y_1 \\ y_2 \\ y_3 \end{bmatrix} = \begin{bmatrix} 40 \\ 30 \\ 0 \\ 0 \\ 0 \end{bmatrix}$$

  Comparing the two solutions:

  o The decision variables $\{t, c\}$ for the primal problem are 0, while the complementary dual variables $\{r_1, r_2\}$ are positive.

  o Similarly, but in the opposite sense, all the slack variables for the primal problem $\{s_1, s_2, s_3\}$ are positive, while all the associated dual variables $\{y_1, y_2, y_3\}$ are 0.

  o Thus, the equality between the objective function's values is also verified.

  $$\mathbf{c.x}^{(1)} = \mathbf{y}^{(1)}.\mathbf{b}$$

  $$\begin{bmatrix} c_t & c_c & c_{s_3} \end{bmatrix}_\mathbf{B} . \mathbf{x_B}^{(1)} = \begin{bmatrix} y_1 & y_2 & y_3 \end{bmatrix}^{(1)} . \mathbf{b}$$

$$[0\ 0\ 0].\begin{bmatrix}60\\80\\30\end{bmatrix}^{(1)}=[0\ 0\ 0]^{(1)}.\begin{bmatrix}60\\80\\30\end{bmatrix}$$

$$0=0$$

In the second tableau, the solution for the primal problem is

$$\begin{bmatrix}t\\c\\s_1\\s_2\\s_3\end{bmatrix}=\begin{bmatrix}30\\0\\0\\20\\30\end{bmatrix}$$

and for the dual problem:

$$\begin{bmatrix}r_1\\r_2\\y_1\\y_2\\y_3\end{bmatrix}=\begin{bmatrix}0\\10\\20\\0\\0\end{bmatrix}$$

Comparing again the two feasible solutions:

o The decision variable, c, for the primal problem is 0, while the complementary dual variable, r_2, is positive; and the decision variable, t, for the primal problem is positive, while the complementary dual variable, r_1, is 0.

o Two slack variables for the primal problem are positive $\{s_2, s_3\}$, while the associated dual variables $\{y_2, y_3\}$ are 0; and while the slack variable for the primal problem, s_1, is 0, the associated dual value, y_1, is positive.

o Thus, equality between the objective function's values is also verified:

$$c.x^{(2)} = y^{(2)}.b$$

$$\left[\begin{array}{ccc} c_t & c_{s_2} & c_{s_3} \end{array}\right]_B . x_B^{(2)} = \left[\begin{array}{ccc} y_1 & y_2 & y_3 \end{array}\right]^{(2)}.b$$

$$\left[\begin{array}{ccc} 40 & 0 & 0 \end{array}\right]. \begin{bmatrix} 30 \\ 20 \\ 30 \end{bmatrix}^{(2)} = \left[\begin{array}{ccc} 20 & 0 & 0 \end{array}\right]^{(2)}. \begin{bmatrix} 60 \\ 80 \\ 30 \end{bmatrix}$$

$$1200 = 1200$$

The complementarity properties can thus be summarized:

- When the primal decision variables are positive, the associated dual values (reduced costs) take 0; and vice versa, when the primal decision variables are 0, the associated dual reduced costs are positive:

$$x_j . r_j = 0$$

- When the primal slack variables are positive, the associated dual variables (shadow costs) take 0; and vice versa, when the primal slack variables are 0, the associated dual values are positive:

$$s_i . y_i = 0$$

- Note that the associated primal-dual variables can be simultaneously 0, indicating the existence of multiple optimal solutions; this can occur either with the pair slack variable-shadow price, (s_i, y_i), or with the pair decision variable-reduced cost, (x_j, r_j).

4.5 DUALITY AND ECONOMIC INTERPRETATION

In general, a primal LP problem that addresses the best resource allocation by maximizing an objective function corresponds to a dual problem minimizing the total value for resources utilization while the related internal prices for each resource are defined.

Note that the objective function value Z for the primal problem is expressed in euros; the coefficients in c are expressed in euros per unit of product, respectively, euros per table and euros per chair; and the primal variables in *x* represent the quantities manufactured for each product, respectively, the quantity of tables and the quantity of chairs, as follows:

$$Z^* = \mathbf{c}.\mathbf{x}^*$$

$$(\text{Euros}) = \left(\frac{\text{Euros}}{\text{Unit of Product}}\right).(\text{Quantity of Product})$$

Similarly, the dual problem's objective function value *W* is expressed in euros; the RHS parameters in *b* represent the available quantities for each resource, respectively, the available quantities of big-blue parts, small-red parts, and the upper-bound on the chairs assembly line. Therefore, the dual variables in *y* are expressed in euros per unit of resource, as follows:

$$W^* = \mathbf{y}^*.\mathbf{b}$$

$$(\text{Euros}) = \left(\frac{\text{Euros}}{\text{Unit of Resource}}\right).(\text{Quantity of Resource})$$

In this way, the dual variables are indicating the unit values per resource, corresponding to an internal point of view associated with the utilization of each resource. The dual variables $\{y_1, y_2, y_3\}$ are expressing the internal values, the dual costs, or the shadow prices for the big-blue parts and the small-red parts, and the availability of the chairs' assembly line.

Thus, the variables $\{y_1, y_2, y_3\}$ also represent the impact on the objective function of unit alterations in the availability of resources. The associated dual values for the optimal solution are (10, 10, 0)* and, in concordance with the primal-dual complementary properties for the optimal solutions, it can be observed that

- The first restriction on big-blue parts is exhausted, all the available parts ($b_1 = 60$) are utilized to produce tables ($t = 20$) and chairs ($c = 20$), then $s_1 = 0$; if one additional big-blue part exists, then one more table can be produced ($t = 21$) with the parts obtained by deconstructing one chair ($c = 19$); the marginal gain will thus be 10, resulting from the difference between increasing one table and reducing one chair ($40 - 30 = 10$), $y_1 = 10$, verifying that $s_1.y_1 = 0$.

- The second restriction on small-red parts is exhausted too; all the available parts ($b_2 = 80$) are utilized to produce tables ($t = 20$) and chairs ($c = 20$), then $s_2 = 0$; if one additional small-red part exists, then one more chair can be produced ($c = 21$) with the parts obtained by deconstructing half a table ($t = 19$); the marginal gain will be 10, resulting from the difference between reducing half a table and adding one chair, $y_2 = -20 + 30 = 10$, and it is also observed that $s_2.y_2 = 0$.

- The third restriction on the chairs assembly is not exhausted. The number of chairs ($c = 20$) does not require full utilization of the available capacity ($b_3 = 30$), and the associated slack variable is positive, $s_3 = 10$; if one unit of the third resource is added, $b_3 = 31$, then the slack variable will increase, $s_3 = 11$, while the optimal solution and the objective function values remain the same; the marginal gain will be 0, $y_3 = 0$, and again $s_3.y_3 = 0$.

Similarly, the dual variables $\{r_1, r_2\}$ express the internal values associated with the primal decision variables $\{t, c\}$, the reduced costs. On one side, these dual variables express the values associated with the primal variables when they enter the solution basis; and when the primal variable is already in the solution basis, i.e., when it is already a basic variable, the complementary reduced cost is 0.

On the other side, $\{r_1, r_2\}$ are excess variables for the dual problem's restrictions; the reduced costs for the production of tables and chairs thus represent the weighted sum of resources utilization, where the weights are the dual costs $\{y_1, y_2, y_3\}$ minus the contributions of such products for the objective function. Namely, for the dual problem:

- In the first restriction, the left-hand side can be seen as the total value of the resources allocated to produce tables:

$$2y_1 + 2y_2 \quad -r_1 \quad = 40$$

The restriction imposes that such value cannot be less than the unit value for tables' gain; if the resources allocated to produce tables overpass the associated gain (40), then r_1 would be positive and a loss occurs, driving the associated primal decision variable to zero, i.e., $t = 0$; in fact, to assume a positive value for the production of tables, the excess variable needs to take zero, $r_1 = 0$, showing that resources are not utilized excessively when compared with the gain for the associated variable, $c_1 = 40$.

- In the second restriction, the first member can be seen as the total value of the resources allocated to produce chairs:

$$1y_1 + 2y_2 + 1y_3 \quad -r_2 = 30$$

This restriction also imposes that such value cannot be less than the unit value for chairs' gain; similarly, if the resources allocated to the production of chairs overpass the associated gain (30), then r_2 would be positive and the associated primal variable becomes zero because a loss is occurring, i.e., $c = 0$; in the same way, to obtain a positive value for the chairs' variable, the associated excess variable is taken as zero, $r_2 = 0$, showing that resources are not utilized excessively when compared with the unit gain for chairs, $c_2 = 30$.

4.6 A FIRST APPROACH TO OPTIMALITY ANALYSIS

Both the duality concepts and the complementary properties for primal-dual solutions provide additional information, and such indications and related attributes are very useful; namely, to evaluate the quality of optimal solutions and the implications of diverse variations.

For example, *What if ...?*

A customer wants to buy tables for a coffee bar in Dublin. The tables are tall requiring three small-red pieces per table. The customer offers to pay the unit price of 45 euros. How should we proceed?

- A suitable procedure would include a third variable for the Dublin tables, d, and integrate this variable in all the LP relations, obtaining the primal model:

$$\max Z = 40t + 30c + 45d$$

subject to

$$\begin{cases} 2t + 1c + 2d \le 60 \\ 2t + 2c + 3d \le 80 \\ \quad 1c \quad \le 30 \end{cases}$$

The solution for this LP will be the same as the prior problem; in fact, there is no sufficient gain in producing the new Dublin table for the unit price, $c_3 = 45$, because one more small-red piece is needed, the second resource would be utilized in excess, and the internal costs need to be well evaluated. In this way, the reduced cost for the third variable would be obtained from

$$2y_1 + 3y_2 \qquad -r_3 = 45$$

For the optimal solution and assuming the internal prices, $y_1 = y_2 = 10$, the Dublin table is utilizing resources worth 50 euros ($2 \times 10 + 3 \times 10 = 50$), then $r_3 = 5$.

The optimal solution for the *What If* primal problem is

$$\begin{bmatrix} t \\ c \\ d \\ s_1 \\ s_2 \\ s_3 \end{bmatrix}^* = \begin{bmatrix} 20 \\ 20 \\ 0 \\ 0 \\ 0 \\ 10 \end{bmatrix}$$

and for the dual problem:

$$\begin{bmatrix} r_1 \\ r_2 \\ r_3 \\ y_1 \\ y_2 \\ y_3 \end{bmatrix}^* = \begin{bmatrix} 0 \\ 0 \\ 5 \\ 10 \\ 10 \\ 0 \end{bmatrix}$$

- Addressing the managers of the furniture factory, a candid suggestion would be to ask the customer to increase the price for the Dublin table by at least 5 euros to balance the total resources utilization. For example, increasing the unit price by 10 euros, and assuming the unit price would be $c_3 = 55$, then the new optimal solution would include the third variable:

$$\begin{bmatrix} t \\ c \\ d \\ s_1 \\ s_2 \\ s_3 \end{bmatrix}^* = \begin{bmatrix} 10 \\ 0 \\ 20 \\ 0 \\ 0 \\ 30 \end{bmatrix}$$

and for the associated dual variables:

$$\begin{bmatrix} r_1 \\ r_2 \\ r_3 \\ y_1 \\ y_2 \\ y_3 \end{bmatrix}^* = \begin{bmatrix} 0 \\ 5 \\ 0 \\ 5 \\ 15 \\ 0 \end{bmatrix}$$

Note that one additional small-red part would allow the production of one additional Dublin table with the parts obtained by disassembling one table; then the objective function would increase by 15 ($1 \times 55 - 1 \times 40$), and $y_2 = 15$.

Also, two additional big-blue parts would allow the production of three additional tables with the parts corresponding to two Dublin tables; then the objective function would increase by 10 ($3 \times 40 - 2 \times 55$), with the unit gain $y_1 = 10/2 = 5$.

4.7 CONCLUDING REMARKS

LP duality and its main concepts are addressed, including primal-dual transformation, the dual Simplex method, the complementary relations of primal and dual solutions, as well as the related economic interpretation.

The introduction of new products into the factory's portfolio is considered, by evaluating resource utilization through the associated dual values. In this way, a first approach to sensitivity analysis is developed; in fact, an initial step on uncertainty treatment is performed by introducing new variables in the LP model.

Additional degrees of freedom are obtained, in close relation with the introduction of primal LP variables, and with relevant enhancement in decision-making. These new primal variables provide an enlarged search space, with one additional dimension per new variable.

Thus, the primal LP problem is complemented with the associated dual problem, and a second decision maker can be integrated in the LP framework. Duality enlarges the results space to the non-feasible region, as well as it is enlarging the alternatives space from the resources utilization point of view. For decision-making purposes, the direct application of two decision makers occurs in Game Theory, as presented in Chapter 8.

In the next chapter, calculus optimization is introduced, as well as Lagrange multipliers and their relation to dual values.

CHAPTER 5

Calculus Optimization

THIS CHAPTER SHOWS THE relations between Lagrange multipliers and linear programming (LP) dual values, while both calculus and linear algebra (LA) tools are used to obtain the optimal solution. Firstly, some basic notions of the differential multivariate calculus are revisited, as a reminder to the general reader of the prerequisite concepts; secondly, a simple optimization problem is presented, namely, the maximization of a two-variable function with one restriction; then, a generalization approach is initiated to consider the maximization of a function with n-variables, while m-restrictions are simultaneously satisfied. The Lagrangean function for the furniture factory problem is built, and the solution is obtained through Cramer's rule. Finally, the Lagrangean solution is compared with the LP-based solution, within the infinitesimal context of the Lagrange multipliers.

5.1 INTRODUCTION

In this section, basic notions of the differential multivariate calculus are revisited, namely, the optimality conditions for one-variable functions, either maxima or minima, and for two-variable functions too. Additionally, the occurrence of an implicit function and "saddle points" are examined, as well as their importance for other developments.

5.1.1 One-Variable Function: Maxima and Minima

Consider $y(x) = x^2$; is $y(0)$ a minimum point for this (one-variable) function?

DOI: 10.1201/9781315200323-5

The first-order derivative $y'(x)$ is

$$\frac{df}{dx} = y'(x) = 2x$$

Note that the derivative $y'(x)$ is positive when the independent variable, x, is greater than zero, and the function $y(x)$ increases when x is positive; when variable x is less than zero, the derivative $y'(x)$ is negative too, and the function $y(x)$ decreases.

The necessary condition for the maxima and minima of a continuous function, with the first derivative taking the value 0, applies:

$$\frac{df}{dx} = 0 \Rightarrow 2x = 0 \Rightarrow x = 0$$

The derivative $y'(x)$ is zero when x takes 0, $y'(0) = 0$, and the derivative signal alters from negative to positive, indicating that the function $y(x)$ alters the decreasing trend to a positive trend. At this point, $y(0)$ no longer decreases and starts to increase, and for that $y(x)$ presents a minimum point when x takes the value 0.

The second-order derivative is positive: the first-order derivative increases for all values of variable x, from negative to zero, and then $y'(x)$ takes positive values. This second-order result indicates positive concavity, that is, the curvature for the function $y(x)$ graph line is upwards (Figure 5.1):

$$\frac{d^2 f}{dx^2} = \left[y'(x) \right]' \Rightarrow (2x)' = 2 \Rightarrow 2 > 0, \quad \forall x$$

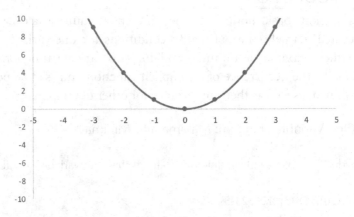

FIGURE 5.1   Minimization for parabola $y = x^2$.

For a maximum point, let us now consider

$$y(x) = -x^2$$

The first-order derivative, $y'(x)$:

$$\frac{df}{dx} = -2x$$

Note that the derivative $y'(x)$ is now positive when the independent variable, x, is less than zero, and the function $y(x)$ increases when x is negative; when variable x is greater than zero, the derivative $y'(x)$ is negative, and the function $y(x)$ decreases.

The necessary condition applies and the first derivative takes the value 0:

$$\frac{df}{dx} = 0 \Rightarrow -2x = 0 \Rightarrow x = 0$$

Again, the derivative $y'(x)$ is zero when x takes 0, $y'(0) = 0$, and the derivative signal alters from positive to negative, indicating that the function $y(x)$ alters the increasing trend to a decreasing one. At this point, $y(x)$ no longer increases and starts to decrease; for that, $y(x)$ presents a maximum point when x takes the value 0.

Thus, note that the second-order derivative is negative: the first-order derivative decreases from positive values to zero, and then $y'(x)$ takes negative values. This second-order result indicates negative concavity, that is, the curvature for the $y(x)$ line is downwards (Figure 5.2):

$$\frac{d^2 f}{dx^2} = [y'(x)]' \Rightarrow (-2x)' = -2 \Rightarrow -2 < 0, \ \forall x$$

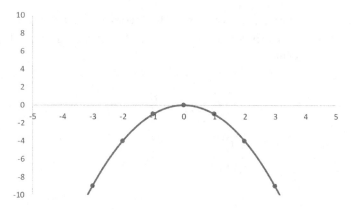

FIGURE 5.2 Maximization for parabola $y = -x^2$.

5.1.2 Two-Variable Function: Maxima and Minima

Consider $f(x, y)$:

$$f(x,y) = x^2 + y^2$$

Is $f(0, 0)$ a maximum?

The first-order (partial) derivatives are as follows:

$$\begin{cases} \dfrac{\partial f}{\partial x} = 2x \\[2mm] \dfrac{\partial f}{\partial y} = 2y \end{cases}$$

The necessary condition for the maxima/minima is also similar to the one-variable function, requiring the first-order partial derivatives to take a value of 0:

$$\begin{cases} \dfrac{\partial f}{\partial x} = 0 \\[2mm] \dfrac{\partial f}{\partial y} = 0 \end{cases} \Rightarrow \begin{cases} 2x = 0 \\ 2y = 0 \end{cases} \Rightarrow \begin{cases} x = 0 \\ y = 0 \end{cases}$$

However, is $f(0, 0)$ a maximum or a minimum?

Note that the partial derivatives of the second order need to be evaluated too, these derivatives corresponding to the elements of the Hessian matrix, respectively:

$$\begin{cases} \dfrac{\partial^2 f}{\partial x^2} = 2; & \dfrac{\partial^2 f}{\partial x \partial y} = 0 \\[2mm] \dfrac{\partial^2 f}{\partial y \partial x} = 0; & \dfrac{\partial^2 f}{\partial y^2} = 2 \end{cases}$$

The sufficient condition is satisfied, the positive values for the second-order derivatives in the Hessian matrix, **H**, indicate positive concavity with an upwards curvature:

$$\frac{\partial^2 f}{\partial x^2} = 2 \ (> 0); \qquad \frac{\partial^2 f}{\partial y^2} = 2 \ (> 0);$$

$$|\mathbf{H}| = 2 \times 2 - 0^2 = 4 \ (> 0)$$

Then, $f(0, 0)$ is a minimum, as shown in the graphic representations in Figure 5.3a (3D view) and Figure 5.3b and c (lateral views).

5.1.3 Two-Variable Function: Saddle Point

Now consider $f(x, y)$:

$$f(x, y) = x^2 - y^2$$

Is $f(0, 0)$ a minimum?

The first-order (partial) derivatives are as follows:

$$\begin{cases} \dfrac{\partial f}{\partial x} = 2x \\[2mm] \dfrac{\partial f}{\partial y} = -2y \end{cases}$$

The necessary condition for the maxima/minima is that the first-order partial derivatives take value 0, then,

$$\begin{cases} \dfrac{\partial f}{\partial x} = 0 \\[2mm] \dfrac{\partial f}{\partial y} = 0 \end{cases} \Rightarrow \begin{cases} 2x = 0 \\ -2y = 0 \end{cases} \Rightarrow \begin{cases} x = 0 \\ y = 0 \end{cases}$$

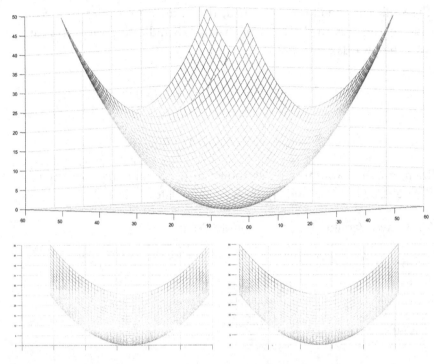

Figure 5.3 (a) Minimization for paraboloid $z = x^2 + y^2$ (perspective view). (b, c) Minimization for paraboloid $z = x^2 + y^2$ (b: lateral view, x-axis; c: lateral view, y-axis).

And now, is $f(0, 0)$ a minimum or a maximum?

Note that the second-order partial derivatives need to be evaluated too:

$$\begin{cases} \dfrac{\partial^2 f}{\partial x^2} = 2; & \dfrac{\partial^2 f}{\partial x \partial y} = 0 \\[3mm] \dfrac{\partial^2 f}{\partial y \partial x} = 0; & \dfrac{\partial^2 f}{\partial y^2} = -2 \end{cases}$$

However, the second-order condition is not satisfied because the partial derivatives indicate different concavities; that is, upward and downward curvatures for x and y, respectively:

$$\frac{\partial^2 f}{\partial x^2} = 2 \;\; (>0); \quad \frac{\partial^2 f}{\partial y^2} = -2 \;\; (<0);$$

$$|\mathbf{H}| = 2 \times (-2) - 0^2 = 4 \;\; (<0)$$

Thus, point $f(0, 0)$ is neither a minimum nor a maximum; it is a *saddle point* as shown in the graphic representation in Figure 5.4a–c.

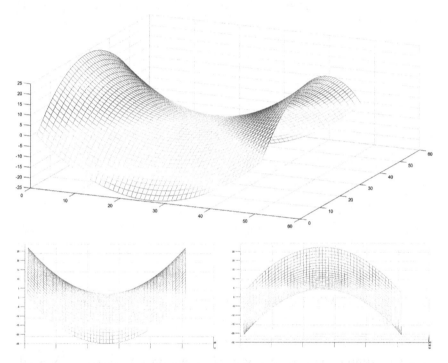

Figure 5.4 (a) Non-optimization for saddle point $z = x^2 - y^2$ (perspective view). (b, c) Non-optimization for saddle point $z = x^2 - y^2$ (b: lateral view, x-axis; c: lateral view, y-axis).

5.1.4 The Implicit Function

Now assume that

$$f(x, y) = x^2 + y^2 = 25$$

When x increases, y is reduced because the sum is constant (equal to 25) and vice versa. The current relation represents all the points (x, y) where the function takes the value of 25, that is, a function isoline. The implicit function can thus be defined as

$$F(x, y) = x^2 + y^2 - 25 = 0$$

The function $F(x, y)$ represents the circumference centered in the origin (0, 0) and the radius is 5. An upper view of the isolines for a bivariate parabola is shown in Figure 5.5.

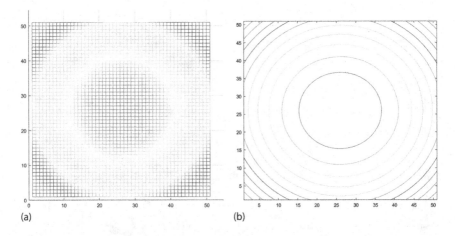

(a)   (b)

FIGURE 5.5   (a,b) Minimization for paraboloid $z = x^2 + y^2$ (a: upper view; b: contour lines).

In this simple case, the one-variable function $y(x)$ can be expressed, that is, it can be explicitly obtained through the two branches (positive and negative) of the square root function:

$$y^2(x) = 25 - x^2$$
$$y(x) = \pm\sqrt{5^2 - x^2}$$

Figure 5.6 shows these two branches, as well as the point (5, 0), since y takes the value 0 when x takes 5.

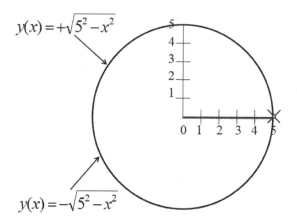

FIGURE 5.6 Implicit function: positive and negative branches, and point (5, 0).

The next set of figures illustrate the implicit relation between y and x: when x diminishes to 4, 3, and 0, then y increases to 3, 4, and 5, respectively. Note that the sum for the squares of the coordinates for all these points—(5, 0), (4, 3), (3, 4), and (0, 5)—is constant and equal to 25. And vice versa, when y diminishes, x increases in such a way as to maintain constant the sum of squares (Figure 5.7a–c).

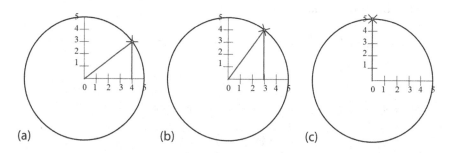

FIGURE 5.7 (a–c) Implicit function, negative relation between variables y and x, while the sum of squares remains constant. (a) Point (4, 3): when x diminishes to 4, y increases to 3. (b) Point (3, 4): when x diminishes to 3, y increases to 4. (c) Point (0, 5): when x diminishes to 0, y increases to 5.

In this case, if the two-variable function $F(x, y)$ is constant, then the total differential is zero:

$$dF(x, y) = 0$$

For that, a relation between the partial derivatives can be obtained, since

$$dF(x,y) = \left(\frac{\partial F}{\partial x}\right)dx + \left(\frac{\partial F}{\partial y}\right)dy = 0$$

and a positive alteration in x provokes a negative impact in y:

$$\left(\frac{\partial F}{\partial y}\right)dy = -\left(\frac{\partial F}{\partial x}\right)dx$$

Finally, the implicit relation $y(x)$ shows that

$$\frac{dy}{dx} = -\frac{\left(\dfrac{\partial F}{\partial x}\right)}{\left(\dfrac{\partial F}{\partial y}\right)}$$

5.2 CONSTRAINED OPTIMIZATION WITH LAGRANGE MULTIPLIERS

The initial case is to find the optima of a two-variable function when submitted to one restriction, that is, one equation.

Now, the problem is

how to obtain the maxima/minima for the function:

$$u = f(x,y)$$

while variables x and y have to satisfy the equation:

$$g(x,y) = 0$$

From the second relation and assuming that $y(x)$ is the implicit function of x, then the derivative of the composite function $u(x, y)$ is

$$\frac{du}{dx} = \frac{\partial f}{\partial x} + \frac{\partial f}{\partial y}\frac{dy}{dx}$$

From the first order necessary condition for optima:

$$\frac{du}{dx} = 0$$

and

$$\frac{\partial f}{\partial x} + \frac{\partial f}{\partial y} \cdot \frac{dy}{dx} = 0$$

From the derivation of the implicit function, $y(x)$:

$$\frac{dy}{dx} = -\frac{\left(\dfrac{\partial g}{\partial x}\right)}{\left(\dfrac{\partial g}{\partial y}\right)}$$

Then, multiplying the partial derivative in the fraction's denominator into the first member:

$$\frac{\partial g}{\partial y} \cdot \frac{dy}{dx} = -\frac{\partial g}{\partial x}$$

and manipulating the negative derivative in the second member to the first member:

$$\frac{\partial g}{\partial x} + \frac{\partial g}{\partial y} \cdot \frac{dy}{dx} = 0$$

Conjugating the two zero equations:

$$\left(\frac{\partial f}{\partial x} + \frac{\partial f}{\partial y} \cdot \frac{dy}{dx}\right) - \lambda \cdot \left(\frac{\partial g}{\partial x} + \frac{\partial g}{\partial y} \cdot \frac{dy}{dx}\right) = 0$$

where λ is known as the Lagrange multiplier.
    Reallocating terms:

$$\left(\frac{\partial f}{\partial x} - \lambda \cdot \frac{\partial g}{\partial x}\right) + \left(\frac{\partial f}{\partial y} - \lambda \cdot \frac{\partial g}{\partial y}\right) \cdot \frac{dy}{dx} = 0$$

The multiplier λ is selected in conformity with

$$\frac{\partial f}{\partial y} - \lambda \cdot \frac{\partial g}{\partial y} = 0$$

and it is also necessary that

$$\frac{\partial f}{\partial x} - \lambda \cdot \frac{\partial g}{\partial x} = 0$$

Thus, in optimal points, the following three equations are simultaneously verified (necessary conditions for the existence of constrained optima):

$$\begin{cases} \dfrac{\partial f}{\partial x} - \lambda \cdot \dfrac{\partial g}{\partial x} = 0 \\[2mm] \dfrac{\partial f}{\partial y} - \lambda \cdot \dfrac{\partial g}{\partial y} = 0 \\[2mm] g(x,y) = 0 \end{cases}$$

Notice that they represent the three partial derivatives in (x, y, λ) of the Lagrangean function:

$$L(x,y,\lambda) = f(x,y) - \lambda \cdot g(x,y)$$

The proposed procedure thus follows:

1. Build the Lagrangean function, $L(x, y, \lambda)$.

2. Set to zero the related first-order partial derivatives.

3. Obtain the values (x, y, λ) that satisfy the system of equations.

5.3 GENERALIZATION OF THE CONSTRAINED OPTIMIZATION CASE

Through the usual generalization approach, the problem of maximizing a n-variables function subject to the simultaneous satisfaction of m-equations is addressed in this sub-section.

And now the problem is

how to obtain the maxima/minima of the function:

$$u = f(x_1, x_2, \ldots, x_n)$$

but the n-variables $(x_1, x_2, \ldots x_n)$ have to satisfy the set of m ($m < n$) equations:

$$\begin{cases} g_1(x_1, x_2, \ldots, x_n) - b_1 = 0 \\ g_2(x_1, x_2, \ldots, x_n) - b_2 = 0 \\ (\ldots) \\ g_m(x_1, x_2, \ldots, x_n) - b_m = 0 \end{cases}$$

The proposed procedure is developed in three steps:

1. Build the Lagrangean function, $L(x, y, \lambda)$:

$$\begin{aligned} L(x_1, x_2, \ldots, x_n, \lambda_1, \lambda_2, \ldots, \lambda_m) &= f(x_1, x_2, \ldots, x_n) \\ &= \lambda_1 \cdot [g_1(x_1, x_2, \ldots, x_n) - b_1] \\ &\quad - \lambda_2 \cdot [g_2(x_1, x_2, \ldots, x_n) - b_2] \\ &\qquad (\ldots) \\ &\quad - \lambda_m \cdot [g_1(x_1, x_2, \ldots, x_n) - b_m] \end{aligned}$$

2. Set to zero the n first-order partial derivatives in the n-variables (x_1, $x_2, \ldots, x_n$):

$$\begin{cases} \dfrac{\partial f}{\partial x_1} - \lambda_1 \cdot \dfrac{\partial g_1}{\partial x_1} - \lambda_2 \cdot \dfrac{\partial g_2}{\partial x_1} - \ldots - \lambda_m \cdot \dfrac{\partial g_m}{\partial x_1} = 0 \\[2mm] \dfrac{\partial f}{\partial x_2} - \lambda_1 \cdot \dfrac{\partial g_1}{\partial x_2} - \lambda_2 \cdot \dfrac{\partial g_2}{\partial x_2} - \ldots - \lambda_m \cdot \dfrac{\partial g_m}{\partial x_2} = 0 \\[2mm] \qquad\qquad (\ldots) \\[2mm] \dfrac{\partial f}{\partial x_n} - \lambda_1 \cdot \dfrac{\partial g_1}{\partial x_n} - \lambda_2 \cdot \dfrac{\partial g_2}{\partial x_n} - \ldots - \lambda_m \cdot \dfrac{\partial g_m}{\partial x_n} = 0 \end{cases}$$

Notice that the first-order partial derivatives in order to the m multipliers ($\lambda_1, \lambda_2, \ldots, \lambda_m$) are driving the m constraints,

$$\begin{cases} -g_1(x_1, x_2, \ldots, x_n) + b_1 = 0 \\ -g_2(x_1, x_2, \ldots, x_n) + b_2 = 0 \\ \quad (\ldots) \\ -g_m(x_1, x_2, \ldots, x_n) + b_m = 0 \end{cases}$$

And finally,

3. Obtain the values set (x, y) that simultaneously satisfies the group of $(n + m)$ equations. That is, solve

$$\begin{cases} \dfrac{\partial L(\mathbf{x}, \lambda)}{\partial x_j} = \dfrac{\partial f(\mathbf{x})}{\partial x_j} - \displaystyle\sum_{i=1}^{m} \lambda_i \cdot \dfrac{\partial g_i(\mathbf{x})}{\partial x_j} = 0, \quad (j = 1, n) \\ \dfrac{\partial L(\mathbf{x}, \lambda)}{\partial \lambda_i} = -g_i(\mathbf{x}) + b_i = 0, \quad (i = 1, m) \end{cases}$$

5.4 LAGRANGE MULTIPLIERS FOR THE FURNITURE FACTORY PROBLEM

The instance treated in Chapter 2, which addressed linear algebra topics, has been selected to illustrate the application of Lagrange multipliers to the furniture factory's problem.

A furniture factory produces tables (*t*) at a profit of 40 euros per table, and chairs (*c*) at a profit of 30 euros per chair (Figure 5.8).

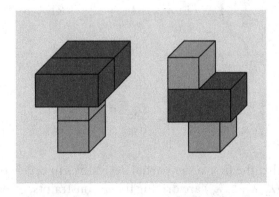

FIGURE 5.8 Furniture factory problem – Tables and chairs.

If the availability of the small-red pieces is 8008 and the availability of the big-blue pieces is 6007, how many tables and chairs need to be built to make the most profit?

Now, the solving procedure uses differential calculus to maximize the profit function, $Z(t, c)$, while satisfying the conditions concerning the availability of the big-blue (B restriction) and small-red (R restriction) components. That is

How should we obtain the maximum value of profit function,

$$Z = 40t + 30c$$

while ensuring that variables t and c satisfy the components availability?

$$\begin{cases} 2t + 1c = 6007 \\ 2t + 2c = 8008 \end{cases}$$

Procedure:

1. Build the Lagrangean function, $L(t, c, \lambda_1, \lambda_2)$.

2. Set to zero the (four) first-order partial derivatives.

3. Obtain the values $(t, c, \lambda_1, \lambda_2)$ that satisfy the system of (four) equations.

Firstly, let us address the Lagrangean function, $L(t, c, \lambda_1, \lambda_2)$:

$$L(t,c,\lambda_1,\lambda_2) = Z(t,c) - \lambda_1 \cdot B(t,c) - \lambda_2 \cdot R(t,c)$$

$$= 40t + 30c - \lambda_1 \cdot [2t + 1c - 6007]$$

$$- \lambda_2 \cdot [2t + 2c - 8008]$$

Secondly, setting to zero the four first-order partial derivatives, and noting that those derivatives related to the multipliers (λ_1, λ_2) are driving the original constraints on the availability of the big-blue (B) and small-red (R) components:

$$
\begin{cases}
\dfrac{\partial L}{\partial t} = 0 \\[6pt]
\dfrac{\partial L}{\partial c} = 0 \\[6pt]
\dfrac{\partial L}{\partial \lambda_1} = 0 \\[6pt]
\dfrac{\partial L}{\partial \lambda_2} = 0
\end{cases}
\Leftrightarrow
\begin{cases}
40 - 2\lambda_1 - 2\lambda_2 = 0 \\
30 - 1\lambda_1 - 2\lambda_2 = 0 \\
-(2t + 1c - 6007) = 0 \\
-(2t + 2c - 8008) = 0
\end{cases}
$$

Finally, let us obtain the values $(t, c, \lambda_1, \lambda_2)$ that simultaneously satisfy the four equations. By reallocating the terms of the system of equations:

$$
\Leftrightarrow
\begin{cases}
2\lambda_1 + 2\lambda_2 = 40 \\
1\lambda_1 + 2\lambda_2 = 30 \\
2t + 1c = 6007 \\
2t + 2c = 8008
\end{cases}
$$

Two subsystems can be observed: variables λ_1, λ_2 are presented only in the first two equations; and variables t, c are presented only in the last two equations. For that, these two subsystems can be solved autonomously, each subsystem considering two variables and two equations.

- Applying Cramer's rule to the first subsystem:

$$
\begin{cases}
2\lambda_1 + 2\lambda_2 = 40 \\
1\lambda_1 + 2\lambda_2 = 30
\end{cases}
$$

then the variables λ_1, λ_2 are directly obtained:

$$
\begin{cases}
\lambda_1 = \dfrac{\begin{vmatrix} 40 & 2 \\ 30 & 2 \end{vmatrix}}{\begin{vmatrix} 2 & 2 \\ 1 & 2 \end{vmatrix}} \\[20pt]
\lambda_2 = \dfrac{\begin{vmatrix} 2 & 40 \\ 1 & 30 \end{vmatrix}}{\begin{vmatrix} 2 & 2 \\ 1 & 2 \end{vmatrix}}
\end{cases}
\Leftrightarrow
\begin{cases}
\lambda_1 = \dfrac{40 \times 2 - 30 \times 2}{2 \times 2 - 1 \times 2} \\[12pt]
\lambda_2 = \dfrac{2 \times 30 - 1 \times 40}{2 \times 2 - 1 \times 2}
\end{cases}
\Leftrightarrow
\begin{cases}
\lambda_1 = \dfrac{80 - 60}{4 - 2} = \dfrac{20}{2} = 10 \\[12pt]
\lambda_2 = \dfrac{60 - 40}{4 - 2} = \dfrac{20}{2} = 10
\end{cases}
$$

- Applying Cramer's rule to the second subsystem:

$$\begin{cases} 2t + 1c = 6007 \\ 2t + 2c = 8008 \end{cases}$$

the decision variables t and c are also obtained:

$$\begin{cases} t = \dfrac{\begin{vmatrix} 6007 & 1 \\ 8008 & 2 \end{vmatrix}}{\begin{vmatrix} 2 & 1 \\ 2 & 2 \end{vmatrix}} \\[2em] c = \dfrac{\begin{vmatrix} 2 & 6007 \\ 2 & 8008 \end{vmatrix}}{\begin{vmatrix} 2 & 1 \\ 2 & 2 \end{vmatrix}} \end{cases} \Leftrightarrow \begin{cases} t = \dfrac{6007 \times 2 - 8008 \times 1}{2 \times 2 - 2 \times 1} \\[1.5em] c = \dfrac{2 \times 8008 - 2 \times 6007}{2 \times 2 - 2 \times 1} \end{cases} \Leftrightarrow \begin{cases} t = \dfrac{12014 - 8008}{4 - 2} = \dfrac{4006}{2} = 2003 \\[1.5em] c = \dfrac{16016 - 12014}{4 - 2} = \dfrac{4002}{2} = 2001 \end{cases}$$

Finally, the optimal solution is

$$\Leftrightarrow \begin{cases} \lambda_1 = 10 \\ \lambda_2 = 10 \\ t = 2003 \\ c = 2001 \end{cases}$$

Comparing the solution obtained with the LP solution, the optimal values for the decision variables, t and c, are confirmed. Note that the Lagrange multipliers and the dual variables for the two linear restrictions on the availability of big-blue and small-red components are equal. Respectively:

$$\frac{\partial L}{\partial x} - \lambda_1 \cdot \frac{\partial B}{\partial x} = 0 \;\Rightarrow\; \lambda_1 = \frac{\partial L}{\partial B}, \text{ and } \lambda_1 = y_1 = 10$$

$$\frac{\partial L}{\partial x} - \lambda_2 \cdot \frac{\partial R}{\partial x} = 0 \;\Rightarrow\; \lambda_2 = \frac{\partial L}{\partial R}, \text{ and } \lambda_2 = y_2 = 10$$

Note that the Lagrange multipliers also represent the partial derivative of the objective function in relation to the associated restriction; then, the dual values can also represent the optimal value's growth per unit variation of the resource availability.

5.5 CONCLUDING REMARKS

Important notions about calculus optimization, with the main purpose to maximize n-variables function subject to the simultaneous satisfaction of m-restrictions, are introduced.

- For that, a generalization approach is initiated with a simple optimization problem, which addresses the maximization of a two-variable function subject to one single restriction; and

- by adding the Lagrange multiplier, which can be perceived as the partial derivative between these two functions, the objective function and the restriction function.

By comparing the Lagrangean solution with the LP solution, it is observed that the Lagrange multipliers and dual values are similar and present a similar interpretation. In fact, a dual value $y = 10$ means that the LP objective function value will increase by 10 when the corresponding RHS parameter increases by 1; note that a derivative in the infinitesimal context corresponds to the curve slope, but the derivative in the LP case is reduced to the constant slope related with a linear function.

Lagrangean relaxation methods are widely disseminated in the scientific literature. They consider the reformulation of some restrictions as new variables in the objective function, which are weighted by the respective dual values or Lagrange multipliers. In this way, the introduction of new variables can be interpreted by the decision maker as an enlargement of the alternatives space.

In addition, solving difficulties are reduced due to the smaller number of restrictions, although counterbalanced with the incorporation of new (dual, shadow) variables. The optimal values for these variables, which are associated with the relaxed restrictions, indicate the optimum set of Lagrange multipliers for the original problem; this property is greatly appreciated by many users of Lagrangean relaxation methods, as well as other decomposition schemes. In fact, Lagrangean methods and decomposition schemes are usually applied in large and complex problems, either by taking advantage of the special structure presented by the mathematical models (e.g., diagonal structures) or by obtaining intermediate results of great interest.

In fact, intermediate results as dual values or Lagrange multipliers allow a better understanding of the various subsystems within the problem at hand, and also represent an enlargement of both the results space and the actions space. Decision makers can use the additional information, the enhanced knowledge about their system and subsystems, to develop robust decision procedures.

Differential calculus and the mathematical conditions for optima, either maxima or minima, are crucial for many exact sciences, technological developments, and engineering applications. Optimization calculus is also key for many developments in economics and enterprise sciences, namely, for Game Theory and multiple types of economic analysis, either static analysis of equilibrium or dynamic analysis with time effects.

CHAPTER 6

Optimality Analysis

Tᴴɪꜱ ᴄʜᴀᴘᴛᴇʀ ᴀᴅᴅʀᴇꜱꜱᴇꜱ ᴛʜᴇ most important procedures for the optimality analysis of a linear programming (LP) problem, namely through a sensitivity analysis of the objective function coefficients and the right-hand side (RHS) parameters, and the introduction of new variables. Based on LP Simplex procedures, a sensitivity analysis is developed in a systematic way, aiming at a simple parametric analysis. The main results of an optimality analysis are the optimality and feasibility ranges for the objective function coefficients and RHS parameters, respectively, and the mapping of alterations onto the optimal solution with the evolution of pertinent coefficients or parameters. Then, the decision maker is provided with new sources of information, and these new insights can enhance the response to external fluctuations in prices, resource availability, technology changes, and many other uncertainty factors.

6.1 INTRODUCTION

In the business landscape, common sense suggests innovating and adding value to products, thereby obtaining a competitive and differentiating advantage for consumers. In this context, the need for mathematical tools is well known, either by modeling the specific problem, by simulating the system behavior, or by optimizing a performance function associated with the system or problem decisions. These techniques are widely used in the academic community, but they are often underestimated, or not properly explored, by managers and decision makers who directly address the situation.

DOI: 10.1201/9781315200323-6

Attention should be paid to a careful post-optimal study of the problem, not only checking results in detail to avoid gross errors, but also carrying out a post-optimal analysis that promotes the full utilization of scarce resources.

- The determination, and subsequent implementation, of a values range for the main indicators or parameters is highly appreciated, usually in the neighborhood of the mathematical optimal point.

- In fact, the values used as parameters or coefficients are typically just estimates, and their static or deterministic nature is not fitting a number of dynamic contexts, for example, strategic purchasing on finance markets, planning of healthcare services, or scheduling of industry operations.

Economic indicators and utility parameters are needed to optimize industrial operations, and can be directly or indirectly obtained through accounting data and operations reports. These data sources constitute new and supplementary elements that support the usefulness of a post-optimal study.

- In industry, they are typically analyzed: profits and losses; purchases and sales, quantities and related prices; operating results; operating costs of each industry process, through material and energy balances, unit efficiencies, and production levels.

- Some of these parameters can be manipulated, since they correspond to internal policies or strategic decisions; therefore, they can be revised if it is considered advantageous.

- The collection of critical data can reveal either the occurrence of bottlenecks or the underutilization of capacities, in addition to the need to alter or update the production system.

Case studies and typical LP problems are often used to find the best way to formulate and solve a specific problem. In addition:

- It is important both to focus each case according to its specific objectives, and to discriminate the model and formulation type. Then, the most appropriate techniques can be selected, and the results set can be obtained.

- Post-optimal studies and real-world applications are frequent in the literature, either addressing large and complex systems or industrial networks that integrate supply, production, and distribution.

- A large set of benefits are collected by large companies that can afford optimization groups; due to their impact in the last decades, post-optimization treatments have now become a common procedure.

Some steps are commonly recommended for the practical resolution of an optimization problem. Namely, it is convenient to

- Describe the real-world situation, detail the specific problem, and consider the associated trade-offs.

- Revisit typical optimization cases and problems, and their attributes.

- Take a look at the framework or background associated with each solution and each method, and verify the implementation steps.

- Develop a mathematical model, and implement adequate procedures, methods, or algorithms to obtain the problem's optimal solution.

However, a weak point in the optimization problems lies in the static nature of the optimal solution. In fact, when the optimal solution is obtained, the best set of decision variables is selected and their values are calculated; however, such optimal values are specifically associated with the input data, the set of parameters that feed the calculations. For that, with a single value assigned to each variable, the values calculated in this way are only valid as long as the instance's data does not change.

Sensitive parameters are those that significantly influence the problem's solution, even when small variations arise. The identification of sensitive parameters is important because they require good data treatment, since these values cannot be modified without causing fluctuations in the optimal solution. Hence, in order to avoid perturbations in the optimal results, it is necessary to accurately define and estimate these parameters. Less sensitive parameters would tolerate some level of uncertainty because they are not as important for the optimal solution.

The analysis of sensitive parameters also allows a better course of action, since a weak initial solution would suggest either data tuning, modeling improvements, or changing the calculation procedures. Therefore, after obtaining the optimal solution, post-optimal studies must be carried out (a sensitivity analysis, a parametric analysis) to identify the problem parameters that are critical or sensitive to the solution's implementation. The purpose is to obtain a solution that retains its optimality in face of data perturbations in the input parameters, as well as the associated range values. For that, it is important to identify the parameters that significantly impact the optimal solution, and their variation range.

6.2 REVISING LP SIMPLEX

In order to better analyze the alterations to the optimal solution due to data perturbations, revisiting the most important LP Simplex procedures is recommended. Namely, important topics to better detail are

- The inverse matrix for the solution basis; as described in linear algebra basics (Chapter 2), the inverse matrix for a linear system of equations allows the treatment of multiple RHS, thus it directly addresses perturbations in the RHS parameters for the LP problem, that is, in the column vector **b**.

  Using the matrix-vector for LP formulation (Chapter 3), the expression to calculate the basic solution is indicated in step 4, in page 58:

  $$\mathbf{x_B}^* = \left(\mathbf{B}^{-1}\right)^* .\mathbf{b}$$

- The matrix form of LP Simplex is very useful, in addition to its coherence with the algebraic form (Chapter 3); note that the Gauss elimination procedures are developed line by line, including all the coefficients on each line, either in the RHS or in the equation's first member, either for the basic variables or the non-basic variables. Gauss elimination procedures also apply to the objective function's line, thus directly addressing perturbations in the objective function's coefficients, that is, in the line vector **c**.

  From the matrix LP procedures (step 6), updating the objective function's line evaluates reduced costs, meaning that it directly evaluates the dual values for non-basic variables:

  $$\mathbf{r} = \mathbf{c_N} - \mathbf{c_B}.\left(\mathbf{B}^{-1}.\mathbf{A_N}\right)^*$$

  $$= \mathbf{c_N} - \mathbf{c_B}.\mathbf{K}^*$$

- Duality topics are also important, such as the evaluation of internal costs or the complementary properties of primal-dual solutions (Chapter 4). These topics are useful in different ways, for example, to directly evaluate the gain (or loss) of producing a new table that utilizes diverse resources. In this way, the activity level of a new product, corresponding to the introduction of a new decision variable of index $(n + 1)$, is directly addressed.

Note that the inverse matrix for the solution basis was saving the Gauss elimination steps that were developed to solve the linear system of equations; in addition, in the problem at hand and in many others of the same type, the solution basis in the first iteration corresponds to the slack variables with the associated identity matrix, I. And similarly for the augmented matrix in the Gauss–Jordan method, after each iteration:

$$\left[\, A \,|\, I \,\right] \leftrightarrow \left[\, I \,|\, A^{-1} \,\right]$$

In the LP tableau, the inverse matrix at each iteration can be found in the coefficients for the slack variables. Additionally, it can also be observed how each equation was obtained from the initial equations, and this observation remains coherent for both the algebraic and the tableau form.

The optimal solution in the final iteration of LP Simplex can be directly obtained from the inverse matrix, B^{-1}, by assuming that the solution basis at hand remains both feasible and optimal. This assumption includes the updating calculations for basic variables and the objective function value, as well as for the shadow prices and reduced costs, and it supports the sensitivity analysis procedure that follows.

6.2.1 Sensitivity Analysis Procedure

Based on the LP Simplex review in face of data perturbations, the following steps are recommended for the sensitivity analysis:

1) Revisit the LP Simplex's final iteration, either in algebraic, tableau, or matrix form.

2) Obtain the inverse matrix for the solution basis.

3) Test the feasibility of the new solution; as described in step 4 of the matrix form, check if the new solution still stands in the feasible region.

4) Test the optimality of the new solution; as described in step 6 of the matrix form, check if the new solution remains optimal.

5) Re-optimize the LP model, if the new solution is not satisfying both step 4 (solution feasibility) and step 6 (solution optimality). For that, update the solution basis and the inverse matrix; this step corresponds to applying Gauss elimination to the new linear system of equations, or to the new LP tableau.

6.3 SENSITIVITY ANALYSIS

In this section, the sensitivity analysis is illustrated through the furniture factory problem, by developing a post-optimal analysis:

i) On the variation in the RHS parameters, $\mathbf{b}'$; a range that remains feasible is then obtained for each RHS parameter.

ii) On the variation in the objective function coefficients, $\mathbf{c}'$; a range that remains optimal is then obtained for each coefficient.

iii) By introducing a new variable.

The LP model under analysis is presented again:

$$\max Z = 40t + 30c$$

subject to

$$\begin{cases} 2t + 1c \leq 60 \\ 2t + 2c \leq 80 \\ 1c \leq 30 \end{cases} \tag{6.1}$$

$$t, c \geq 0$$

And the optimal tableau, from the LP Simplex's final iteration (Section 3.3), is again presented in Table 6.1.

The inverse matrix for the solution basis $\{t, c, s_3\}$ can easily be obtained, or calculated. In addition, using the Gauss–Jordan method, the inverse matrix can be observed in the three columns associated with the slack variables in the LP tableau, namely:

TABLE 6.1   Final Tableau with the Optimal Solution

Basis	t	c	s_1	s_2	s_3	RHS
t	1	0	1	−0.5	0	20
c	0	1	−1	1	0	20
s_3	0	0	1	−1	1	10
Z	0	0	−10	−10	0	**−1400**

$$\left(\mathbf{B}^{-1}\right)^* = \begin{bmatrix} 1 & -\tfrac{1}{2} & 0 \\ -1 & 1 & 0 \\ 1 & -1 & 1 \end{bmatrix}$$

6.3.1 Variation in the RHS Parameters, b′

Data perturbation in the RHS parameters would make the new solution non-feasible, in the case of at least one restriction is not satisfied, and then a basic variable presents negative value. Due to the non-negativity property of LP variables, there is no change to the solution basis if the new solution satisfies the *feasibility test*:

$$\mathbf{x_b}^* = \left(\mathbf{B}^{-1}\right)^* .\mathbf{b}' \geq 0$$

then

$$Z' = \mathbf{y}^* .\mathbf{b}'$$

For the current instance, the range to remain feasible for each RHS parameter, as well as the allowable increase or decrease, is obtained from

$$\mathbf{b}^* = \begin{bmatrix} 1 & -\tfrac{1}{2} & 0 \\ -1 & 1 & 0 \\ 1 & -1 & 1 \end{bmatrix} . \begin{bmatrix} b_1 \\ b_2 \\ b_3 \end{bmatrix} \geq 0$$

and

$$
\left\{
\begin{array}{l}
b_1 - \dfrac{b_2}{2} + 0 b_3 \geq 0 \\[2mm]
-b_1 + b_2 + 0 b_3 \geq 0 \\[2mm]
b_1 - b_2 + b_3 \geq 0
\end{array}
\right.
\Leftrightarrow
\left\{
\begin{array}{l}
b_1 \geq \dfrac{b_2}{2} \\[2mm]
b_1 \leq b_2 \\[2mm]
b_1 - b_2 + b_3 \geq 0
\end{array}
\right.
$$

A *set paribus* analysis follows:

- From $\begin{cases} b_2 = 80 \\ b_3 = 30 \end{cases}$

    then

$$
\left\{
\begin{array}{l}
b_1 \geq \dfrac{80}{2} \\[2mm]
b_1 \leq 80 \\[2mm]
b_1 - 80 + 30 \geq 0
\end{array}
\right.
\Leftrightarrow
\left\{
\begin{array}{l}
b_1 \geq 40 \\[2mm]
b_1 \leq 80 \\[2mm]
b_1 \geq 50
\end{array}
\right.
$$

Since $b_1 = 60$ and the range to remain feasible is [50, 80] then

- The allowable decrease for b_1 is 10 (from 60 to 50).

- The allowable increase for b_1 is 20 (from 60 to 80).

    o From $\begin{cases} b_1 = 60 \\ b_3 = 30 \end{cases}$
       then

$$
\left\{
\begin{array}{l}
60 \geq \dfrac{b_2}{2} \\[2mm]
60 \leq b_2 \\[2mm]
60 - b_2 + 30 \geq 0
\end{array}
\right.
\Leftrightarrow
\left\{
\begin{array}{l}
b_2 \leq 120 \\[2mm]
b_2 \geq 60 \\[2mm]
b_2 \leq 90
\end{array}
\right.
$$

Since $b_2 = 80$ and the range to remain feasible is [60, 90] then

- The allowable decrease for b_2 is 20 (from 80 to 60).
- The allowable increase for b_2 is 10 (from 80 to 90).

o From $\begin{cases} b_1 = 60 \\ b_2 = 80 \end{cases}$

then

$$\begin{cases} 60 \geq \dfrac{80}{2} \\ 60 \leq 80 \\ 60 - 80 + b_3 \geq 0 \end{cases} \quad \Leftrightarrow \quad \begin{cases} 60 \geq 40 \\ 60 \leq 80 \\ b_3 \geq 20 \end{cases}$$

Since $b_3 = 30$ and the range to remain feasible is $[20, +\infty[$ then

- The allowable decrease for b_3 is 10 (from 30 to 20).
- The allowable increase for b_3 is unbounded (from 30 to $+\infty$).

6.3.2 Variation in the Objective Function Coefficients, c′

Data alterations in the objective function coefficients would make the new solution non-optimal, in which case the optimality condition is not satisfied. That is, the optimal basic solution cannot be improved by entering a non-basic variable, and there is no change in the solution basis if the new solution satisfies the *optimality test*:

$$r' = c_N' - c_B' \cdot \left(B^{-1} . A_N \right)^*, \quad \text{verifies} \quad \begin{cases} \geq 0, & \text{minimization} \\ \leq 0, & \text{maximization} \end{cases}$$

then

$$Z' = c_B' x_b^*$$

In this instance, the basic variables are $\{t, c, s_3\}$ and the non-basic variables are $\{s_1, s_2\}$; the objective function coefficients are the associated coefficients. The non-basic matrix A_N can be rapidly updated to matrix K using the inverse matrix, or observing the columns for the non-basic variables

$\{s_1, s_2\}$ within the optimal solution in Table 6.1, or even revisiting step 5 in the final iteration of Section 3.4. Thereafter,

$$\mathbf{K}^* = \left(\mathbf{B}^{-1}.\mathbf{A_N}\right)^* = \begin{bmatrix} 1 & -0.5 \\ -1 & 1 \\ 1 & -1 \end{bmatrix}$$

For the current instance, the range to remain optimal for each objective function coefficient, as well as the allowable increase or decrease, can be obtained from

$$\mathbf{r}' = \begin{bmatrix} 0 & 0 \end{bmatrix} - \begin{bmatrix} c_1 & c_2 & 0 \end{bmatrix} . \begin{bmatrix} 1 & -\frac{1}{2} \\ -1 & 1 \\ 1 & -1 \end{bmatrix} \le 0$$

then

$$\begin{cases} -(c_1 - c_2 + 0) \le 0 \\ -\left(-\dfrac{c_1}{2} + c_2 - 0\right) \le 0 \end{cases} \Leftrightarrow \begin{cases} c_1 - c_2 \ge 0 \\ -\dfrac{c_1}{2} + c_2 \ge 0 \end{cases} \Leftrightarrow \begin{cases} c_1 \ge c_2 \\ c_1 \le 2c_2 \end{cases}$$

A *set paribus* analysis follows:

- In the current instance, $c_2 = 30$ and

$$\begin{cases} c_1 \ge 30 \\ c_1 \le 2 \times 30 \end{cases} \Leftrightarrow \begin{cases} c_1 \ge 30 \\ c_1 \le 60 \end{cases}$$

Since $c_1 = 40$ and the range to remain optimal is [30, 60], then

- The allowable decrease for c_1 is 10 (from 40 to 30).

- The allowable increase for c_1 is 20 (from 40 to 60)

   o At the current instance, $c_1 = 40$ and

$$\begin{cases} 40 \ge c_2 \\ 40 \le 2c_2 \end{cases} \Leftrightarrow \begin{cases} c_2 \le 40 \\ c_2 \ge 20 \end{cases}$$

Since $c_2 = 30$ and the range to remain optimal is [20, 40], then

- The allowable decrease for c_2 is 10 (from 30 to 20).
- The allowable increase for c_2 is 10 (from 30 to 40).

6.3.3 Introduction of New Variables

Introducing a new variable, index $(n + 1)$, the basic solution remains optimal if

$$r_{n+1} = c_{n+1} - \mathbf{c_B} \cdot \left(\mathbf{B^{-1}}\right)^* \cdot \mathbf{a}_{n+1}, \quad \text{verifies} \quad \begin{cases} \geq 0, & \text{minimization} \\ \leq 0, & \text{maximization} \end{cases}$$

and

$$\mathbf{a_{n+1}}^* = \left(\mathbf{B^{-1}}\right)^* \cdot \mathbf{a}_{n+1}$$

For the current case, to remain optimal the range is obtained from

$$r_{n+1} = c_{n+1} - \begin{bmatrix} 40 & 30 & 0 \end{bmatrix} \cdot \begin{bmatrix} 1 & -\frac{1}{2} & 0 \\ -1 & 1 & 0 \\ 1 & -1 & 1 \end{bmatrix} \cdot \begin{bmatrix} a_{1,n+1} \\ a_{2,n+1} \\ a_{3,n+1} \end{bmatrix}$$

$$= c_{n+1} - \begin{bmatrix} 10 & 10 & 0 \end{bmatrix} \cdot \begin{bmatrix} a_{1,n+1} \\ a_{2,n+1} \\ a_{3,n+1} \end{bmatrix} \leq 0$$

Let it be the Dublin table (in Section 4.6, page 91), which requires two big-blue parts and three small-red parts, with the customer offering to pay 45 euros per Dublin table. The calculations for the new variable, with index $(n+1)$, follow.

From

$$\mathbf{a}_{n+1} = \begin{bmatrix} 2 \\ 3 \\ 0 \end{bmatrix}$$

then

$$c_{n+1} - \begin{bmatrix} 10 & 10 & 0 \end{bmatrix} . \begin{bmatrix} 2 \\ 3 \\ 0 \end{bmatrix} \leq 0$$

$$c_{n+1} - 50 \leq 0 \Leftrightarrow c_{n+1} \leq 50$$

The current solution basis is still optimal, since the price offered for the Dublin table is below 50.

With $c_{n+1} = 50$, or larger, the new variable enters the solution basis and the activity level for the new Dublin table is positive!

A brief comparison can be made with the *What if…* example for duality in Chapter 4; the utilization of two units from the first resource (big-blue parts) with an internal cost of 10, and three units from the second resource (small-red parts) also with an internal cost of 10, will result in a total resources consumption of 50 to produce the Dublin table. Then, a price of 45 euros per Dublin table would not be appealing enough to produce it, while a price of 55 euros would drive an improvement on the objective function of 100, due to the production of 20 Dublin tables times the marginal value 5 (55 – 50).

6.3.4 Variation in Non-Basic Coefficients

In the case of a variation in the non-basic coefficients, the LP model is updated as in the former situation, similarly as if a new variable is introduced in c_N or A_N. The Gauss elimination procedure is performed and it corresponds to the matrix-vector operations within the optimal solution basis.

Then, the new solution remains optimal if

$$r' = c'_N - c_B . \left[\left(B^{-1} \right)^* . A'_N \right], \quad \text{verifies} \quad \begin{cases} \geq 0, & \text{minimization} \\ \leq 0, & \text{maximization} \end{cases}$$

6.3.5 Variation in Basic Coefficients

In the case of a variation in the basic coefficients, the LP model is updated as in the former situations, and the new values are introduced in c_B or B. Note that the inverse matrix B^{-1} needs to be updated, and the optimal solution can present different values, that is, different activity levels between the basic variables.

The solution basis remains unchanged if the new solution satisfies both the feasibility and the optimality tests:

$$\begin{cases} \mathbf{r'} = \mathbf{c_N} - \mathbf{c_B}.(\mathbf{B^{-1}}.\mathbf{A_N}) \quad , \qquad \text{verifies} \quad \begin{cases} \geq 0, \quad \text{minimization} \\ \\ \leq 0, \quad \text{maximization} \end{cases} \\ \\ \mathbf{b^*} = \mathbf{B^{-1}}.\mathbf{b} \geq 0 \end{cases}$$

6.3.6 Introduction of New Restriction

When introducing a new restriction, it must be directly checked if the optimal solution does or does not satisfy the new restriction.

- If positive, obviously, the optimal solution remains feasible.

- If negative, a suitable alternative is to re-optimize using the dual Simplex algorithm.

6.4 PARAMETRIC ANALYSIS

In this section, the parametric analysis is illustrated through the furniture factory problem, by developing a post-optimal analysis:

i) Mapping the variation of the RHS parameter, b_1; a course of action for the optimal production is associated with the parameter under analysis, while the evolution of both the dual value y_1 and the objective function value Z is compared.

ii) Mapping the variation of the objective function coefficient, c_1; a course of action for the optimal production is also associated with the coefficient at hand, while the evolution of the decision variable t and the objective function value Z is compared.

6.4.1 Parametric Analysis: b_1

Now the LP model considers the parameterization of the RHS value, b_1, and the optimal solution values are obtained with b_1 evolving in the range 0–200.

$$\max Z = 40t + 30c$$

subject to

$$\begin{cases} 2t + 1c \leq b_1 \\ \\ 2t + 2c \leq 80 \\ \\ 1c \leq 30 \end{cases} \qquad (6.2)$$

$$t, c \geq 0$$

The synopsis table, including the solution basis associated with the variation of the parameter b_1, follows.

TABLE 6.2   Evolution of Dual Value y_1, Objective Function Value Z, and Basic Variables with the Variation of Parameter b_1

b_1-Blue	y_1-Dual	Z^*	Enter Var	Depart Var
0	30	0	s_1, s_2, s_3	s_1
1	30	30	c	
10	30	300		
20	30	600		
29	30	870		
30	30	900		s_3
30	20	900	t	
31	20	920		
40	20	1100		
49	20	1280		
50	20	1300		s_2
50	10	1300	s_3	
51	10	1310		
60	10	1400		
70	10	1500		
79	10	1590		
80	10	1600		c
80	0	1600	s_1	
81	0	1600		
90	0	1600		
100	0	1600		
110	0	1600		
120	0	1600		
130	0	1600		
140	0	1600		
150	0	1600		
160	0	1600		
170	0	1600		
180	0	1600		
190	0	1600		
200	0	1600		

For the parameter b_1 under analysis, in the range 0–200 for the first restriction, Table 6.2 shows the evolution of adding big-blue parts to the first resource.

- Commencing with $b_1 = 0$, the trivial solution with $Z = 0$ follows, since the furniture's production cannot start without the first resource, and $t = c = 0$; the decision variables are non-basic variables, and the slack variables are assumed as basic variables.

  Note that for $s_1 = 0$, a basic variable takes 0, and a degenerate solution is obtained.

- With $b_1 = 1$, when adding one big-blue part, up to 30, the associated dual value is $y_1 = 30$; that is, the objective function value increases by 30 per unit of the first resource, because one chair can be produced by adding one big-blue part; proportionally, 10 parts allow the production of 10 chairs, and 30 parts allow the production of 30 chairs, thus obtaining a total gain of $Z = 900$.

  Hereafter, the third resource is exhausted, the capacity of the chairs' assembly is fully utilized, and $s_3 = 0$; this slack variable then exits the solution basis, while the alternative of producing tables is selected, and decision variable t enters the basis.

- When adding other big-blue parts and b_1 within the range 30–50, the associated dual value is $y_1 = 20$ and variable t is positive; that is, the objective function value increases by 20 per unit of the first resource, because one table can be produced by adding two big-blue parts; proportionally, 20 additional parts allow the production of 10 tables, with a marginal gain of 400 that drives a total gain of $Z = 1300$.

  Hereafter, the second resource is exhausted, the small-red parts are fully consumed, and $s_2 = 0$; this slack variable then exits the solution basis, while the alternative of producing one table with the parts obtained from deconstructing one chair is selected; in this way, the number of chairs diminishes, and the slack variable s_3 re-enters the basis.

- Still adding big-blue parts and b_1 within the range 50–80, the dual value is $y_1 = 10$ and the objective function value increases by 10 per unit of the first resource; this occurs because one table is produced by adding one big-blue part and deconstructing one chair to obtain small-red parts, with the corresponding marginal gain, $40 - 30 = 10$; proportionally, 30 big-blue additional parts allow a marginal gain of 300, which drives a total gain of $Z = 1600$.

At this point, the existing chairs are all deconstructed, and $c = 0$; this decision variable exits the solution basis, and the slack variable s_1 enters the solution basis; this means that additional big-blue parts will not be fully utilized.

- By adding other big-blue parts and b_1 greater than (or equal to) 80, the dual value is $y_1 = 0$, and the total gain remains constant, $Z = 1600$; the objective function value does not increase by additional units of the first resource, because adding big-blue parts without additional small-red parts ($s_2 = 0$) does not allow the production of additional furniture, either tables or chairs.

From this point on, with a larger number of big-blue parts and slackness in the first resource, the best alternative is to produce tables until the second resource is exhausted; the number of small-red parts is assumed constant, 80, and this upper limit keeps constant the objective function value.

The evolution of the objective function value Z with parameter b_1's variation is presented in Figure 6.1. The Z values obviously augment with the number of big-blue parts, but the increasing rate of Z's line successively diminishes; finally, a plane occurs, that is, a horizontal line segment corresponds to a constant value for Z.

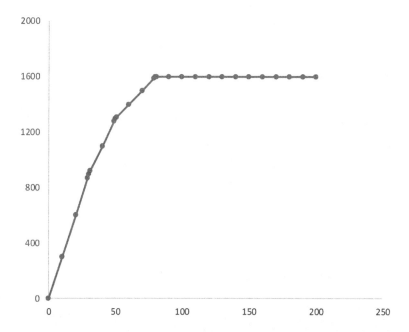

FIGURE 6.1   Variation of the objective function value Z with parameter b_1.

The inclination of the Z line successively diminishes, that is, the impact of the additional units of the first resource is reduced. This corresponds to a successive reduction in the dual value y_1 for the first restriction, as indicated in Chapter 5, where Lagrange multipliers were presented. In the calculus framework, dual value y_1 corresponds to the Lagrange multiplier for the first restriction, λ_1, which leads to the partial derivative of the objective function in relation to the first restriction, that is

$$\frac{\partial f}{\partial x} - \lambda_1 . \frac{\partial g_1}{\partial x} = 0 \Rightarrow \lambda_1 = \frac{\partial f}{\partial g_1}$$

The variation in the dual value y_1 for the first restriction, which is directly associated with the parameter b_1 under analysis, is presented in Figure 6.2. A descending stepwise function is observed, with the levels 30, 20, 10, and 0, corresponding to the marginal gains of additional big-blue parts, as described previously in this section.

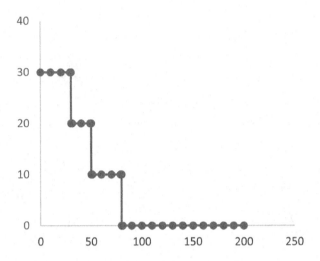

FIGURE 6.2 Variation of the dual value y_1 with parameter b_1.

Using a common-sense approach: (i) the dual value y_1 presents high values when the resource is available in low quantities only, when it is a limited resource of high importance; (ii) the resource's value successively diminishes when the relative quantities increase, when the resource becomes increasingly available; and (iii) the resource's marginal value is null when the resource is available in high quantities, as it is not as relevant as other resources that actively act as limiting resources and bound the optimal solution.

6.4.2 Parametric Analysis: c_1

Now, the LP model considers the parameterization of the objective function coefficient, c_1, and the optimal solution values are obtained with c_1 evolving in the range –10 to 200.

$$\max Z = c_1 t + 30c$$

subject to

$$\begin{cases} 2t + 1c \le 60 \\ 2t + 2c \le 80 \\ 1c \le 30 \end{cases} \tag{6.3}$$

$$t, c \ge 0$$

The synopsis table, including the solution basis (entering variable and departing variable), for the variation of parameter c_1, follows (Table 6.3).

TABLE 6.3 Evolution of Decision Variable t, Objective Function Value Z, and Basic Variables with the Variation of Parameter c_1

c_1	Tables	Z^*	Enter Var	Depart Var
−10	0	900	c, s_1, s_2	
−1	0	900		
0	0	900		s_2
0	10	900	t	
1	10	910		
10	10	1000		
20	10	1100		
29	10	1190		
30	10	1200		s_1
30	20	1200	s_3	
31	20	1220		
40	20	1400		
50	20	1600		
59	20	1780		
60	20	1800		c
60	30	1800	s_2	
61	30	1830		
70	30	2100		
80	30	2400		
90	30	2700		
100	30	3000		
110	30	3300		
120	30	3600		
130	30	3900		
140	30	4200		
150	30	4500		
160	30	4800		
170	30	5100		
180	30	5400		
190	30	5700		
200	30	6000		

For the coefficient c_1 under analysis, in the range −10 to 200 for the first decision variable, t, Table 6.3 shows the evolution of increasing the unit gain for producing tables.

- Commencing with a negative unit gain, $c_1 \leq 0$, the optimal solution assumes $t = 0$, since the production of tables would lead to a marginal loss in such a case; however, in the case of existing orders for tables, providing good service to clients and excellent customer relations could originate from this type of negative gain.

  The entire production system is focused on producing chairs ($c_2 = 30$), with the activity level bounded by the third restriction ($b_3 = 30$), leading to a total $Z = 30 \times 30 = 900$; while the third slack variable is non-basic ($s_3 = 0$), the two other restrictions are not active and the related slack variables, $s_1 = 30$ and $s_2 = 20$, belong to the solution basis.

- With $c_1 = 1$, when increasing the unit gain to 30 euros, the associated activity level for producing tables is $t = 10$; that is, the objective function value increases by 10 per euro of first activity, because 10 tables can be produced with the remaining 20 small-red parts; proportionally, a unit gain of 10 euros allows a partial gain of 100, and 30 euros allows a partial gain of 300, thus obtaining a total gain of $Z = 1200$.

  Note that the furniture factory is still focused on producing chairs ($c_2 = 30$), the activity level for tables is just utilizing the resources that are still available, either big-blue parts or small-red parts.

  In this range, the second resource of small-red parts was exhausted, $s_2 = 0$, due to starting the production of tables; then, the slack variable s_2 exited the solution basis, while the decision variable t entered the basis.

- When increasing the unit gain with c_1 in the range 30–60, the associated activity level is $t = 20$; that is, the objective function value increases by 20 per unit gain of first activity, because another 10 tables can be produced with the remaining 20 big-blue parts; proportionally, increasing the unit gain by 10 euros allows a partial gain of 200, and increasing the unit gain by 30 euros allows a partial gain of 600, thus obtaining a total gain of $Z = 1800$.

Now, the furniture factory is producing a mix of tables and chairs ($t = c = 20$), the activity level for producing tables is increasing by using the parts made available by reducing the production of chairs.

In this range, the first resource of the big-blue parts was exhausted, $s_1 = 0$, and the slack variable s_1 exited the solution basis; by reducing the activity level for producing chairs, the slack variable, s_3, was positive and thus entered the solution basis.

- When the unit gain c_1 is greater than (or equal to) 60, the associated activity level is $t = 30$; that is, the objective function value increases by 30 per unit gain of first activity, because another 10 tables can be produced by reducing the activity level of producing chairs to zero; proportionally, increasing the unit gain to 100 euros allows a total gain of $Z = 3000$, and increasing the unit gain to 200 euros allows a total gain of $Z = 6000$.

Now, the furniture factory is focused on producing tables ($t = 30$), the activity level is bounded by the first restriction ($b_1 = 60$), and the first slack variable is non-basic ($s_1 = 0$); the two other restrictions are not active and the related slack variables, $s_2 = 20$ and $s_3 = 30$, belong again to the solution basis.

From now on, the best alternative is to produce only tables until the first resource is exhausted; the number of big-blue parts is assumed constant, 60, and this upper limit keeps constant the activity level for producing tables ($t = 30$), and the increasing rate for the objective function value.

The evolution of the objective function value Z with parameter c_1's variation is presented in Figure 6.3. The Z values obviously augment with the unit gain for tables, and the increasing rate of the Z's line is successively augmented; finally, a line segment with a constant inclination corresponds to a constant increase ($t = 30$) for Z.

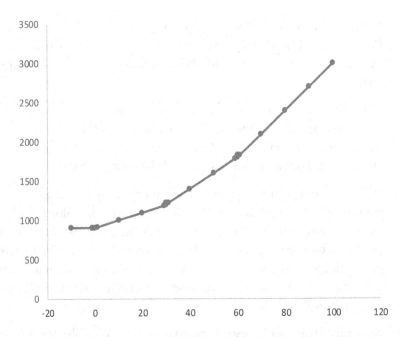

FIGURE 6.3 Variation of the objective function value Z with parameter c_1.

The inclination of the Z line successively augments, that is, the impact of additional unit gains of the first variable becomes increasingly important. This corresponds to a successive increase in the optimal quantities for the first decision variable; in the calculus framework, it corresponds to the partial derivative of the objective function in relation to coefficient c_1 for the first decision variable, that is, the quantity of tables t:

$$t = \frac{\partial f}{\partial c_1}$$

The variation in the quantity of tables, t, for the first variable, which is directly associated with the parameter c_1 under analysis, is presented in Figure 6.4. Now, an ascending stepwise function is observed, with levels 0, 10, 20, and 30, corresponding to the optimal quantities of tables, as described previously.

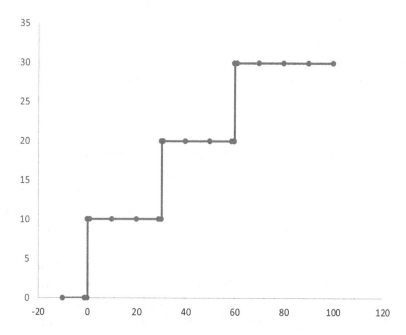

FIGURE 6.4   Variation of the decision variable t with parameter c_1.

Again, using a common-sense approach: (i) the activity level for the first variable t is null when the unit gain is less than or equal to 0; (ii) the activity level for producing tables is successively augmented when the unit gain increases, because the price becomes increasingly appealing; and (iii) the activity level t presents the highest value for producing tables, the highest feasible value that corresponds to a constant quantity of tables represented in the horizontal line segment.

6.5 CONCLUDING REMARKS

Important procedures for LP optimality analysis are addressed in this chapter, in order to cope either with the utilization of deterministic data, the external factors impacting product prices or the introduction of new products, or the availability of scarce resources. A first approach to uncertainty is outlined, and the deterministic treatment of coefficients and parameters is enlarged; in fact, their variability is considered through parametric analysis, and the most sensitive data for the optimal solution is also analyzed.

Based on the matrix form of LP Simplex, a sensitivity analysis is developed in a systematic way, addressing both the objective function coefficients and the RHS parameters; the main results are the optimality ranges in the first case, and the feasibility ranges in the latter. In this way, decision makers have access to additional information that allows a better response to external fluctuations, namely, by addressing uncertainty either in product prices or in resources availability.

When introducing new decision variables, the alternatives space is also enlarged, with new products and the corresponding production levels. In this case, the comparison of alternatives still considers one single attribute, for example, the maximum gain for the production levels associated with the optimal solution (or the minimum cost associated with the optimal allocation of resources).

A simple parametric analysis is also developed, by mapping the alterations to the optimal solution with the evolution of objective function coefficients and RHS parameters. Additionally, multiple optimal solutions are identified that bring new possibilities for the decision maker who manages production operations; the alternatives space can also include linear combinations of these multiple solutions and still present the LP optimal value. In this case, the binary comparison of alternatives needs to surpass the single attribute frame, and to properly consider other attributes related to the equivalence, indifference, or preference between alternatives, or even to outline a preferences relation system.

However, some limitations to the LP modeling are also verified, namely:

- The occurrence of non-integer LP solutions; such feasible solutions have no corresponding actions or activities when translated to the real world; for example, when producing an integer number of tables, constructing an integer number of houses, to buy or sell a finance asset, or by answering yes/no to a new plant location or expansion.

- The proportionality assumption is relevant for LP modeling; however, proportionality between production costs and production levels sometimes does not hold; for example, even if the production level is zero or the manufacturing line is stopped, several fixed charge costs associated with maintenance and materials holding are incurred; in addition, LP dual costs associated with resources availability increase when the resource is scarce, and decrease when the resource is available in large quantities.

The consideration of integer variables in LP drives integer linear programming (ILP), which constrains the LP framework to obtain feasible solutions that are also possible for real-world implementation. Additionally, ILP allows better modeling by including logic and semi-logic relations. These ILP-based qualitative improvements are obtained by constraining the results space for decision variables, but they come at a cost to quantitative bounding, as presented in Chapter 7.

Integer Linear Programming

THIS CHAPTER INTRODUCES THE important topic of integer linear programming (ILP), which provides (i) qualitative improvements due to more realistic solutions, namely, when decision variables require integer values; and (ii) better modeling capabilities, for example, with binary variables formulating contingency decisions or fixed charge costs. However, such qualitative improvements come at a cost of quantitative reductions: the LP optimal value is the hard limit for the branch-and-bound (B&B) method, also considering that the solutions space is successively reduced and the objective value is successively cut. In the opposite sense to the generalization approach, B&B follows a reduction approach that quantitatively constrains the objective function; however, both the modeling features and the integrity attributes largely enhance decision-making through ILP.

7.1 INTRODUCTION

The properties of linear programming (LP) are widely known, namely those that facilitate sensitivity analysis, including the study of variations in the objective function coefficients or in the right-hand side (RHS) parameters.

DOI: 10.1201/9781315200323-7

However, decision variables are constrained to take integer values in a diversity of LP problems. For instance:

- The decision variables will represent integer quantities; for example, the production levels for tables and chairs in the furniture factory problem.

- In fact, the divisibility assumption in LP does not hold in many real-world contexts; in the case of a fractional solution for tables and chairs, such a solution cannot be directly deployed in the real world, while some kind of interpretation can be related to variables' continuity.

- In some cases, the LP proportionality assumption does not fit well with real-world needs; for example, when addressing batch production in manufacturing industries, setup or cleanup costs can occur once the production equipment starts; for that, non-linear or logic-based formulations are commonly required to model fixed charge costs and scale economies.

In a simple way, the ILP problem corresponds to the LP problem complemented with integrity restrictions; that is, some or all of the decision variables must present integer values. When this additional restriction of integrity only has to be observed by some but not all variables, then mixed integer linear programming (MILP) occurs. For example, in MILP problems, binary decision variables are associated with plant and warehouse location, equipment selection, as well as the respective sizing and production scheduling, while the remaining variables (e.g., production quantities, fluxes, sales, purchases) still take non-integer values.

As mentioned, MILP addresses integer variables in LP problems, allowing more realistic models by adapting the integer solution to the physical context, or even inducing logical-based decisions ("true" or "false") when such variables are restricted to binary values (1 or 0). Thus, each of these binary variables is associated with a logical attribute, whose meaning gives rise to a decision ("yes" or "no") in direct correspondence with the ILP optimal solution. These problems, where all variables are of this binary type, are also known as binary integer programming (BIP) problems.

In fact, one of the main fields of application of integer variables corresponds to the more limiting BIP problems. However, integrity restrictions cause significant difficulties in the calculation of ILP solutions, either MILP or BIP. The branch-and-bound method typically requires two LP solutions at each iteration, by generating two branches associated with the binary variable (values: 0 or 1) or the integer variable (values: k or $k + 1$) at hand; then, the B&B method adds new restrictions, bounding the search space at each step, until the search is complete and no further improvements to the binary or integer solution can be made.

Thus, ILP research has undergone a rapid evolution because of the importance of accurately solving ILP problems; additionally, the significant developments in ILP solution methods (e.g., branch-and-cut; cutting planes; decomposition; heuristics; reformulations; and variable redefinitions) are also making the formulation of large-dimension ILP problems more attractive.

7.2 SOLVING INTEGER LINEAR PROGRAMMING PROBLEMS

Although the B&B method can present several specificities, either when solving a BIP or a ILP, it can be summarized in the following pair of fundamental steps:

- **Branching:** Consists of solving the father problem, by relaxing the integrity restrictions; then, creating a pair of LP sub-problems of smaller dimensions, which are obtained by fixing the value k for one of the integer variables (e.g., the integer part obtained for that variable, $k = \text{INT}(x)$, in one case; and $k = \text{INT}(x) + 1$, in the other case). The strategy associated with the selection both of the sub-problem and the variables to be analyzed largely depends on the type of problem at hand; in order to achieve an efficient calculation, such strategy should be well aligned with the most important decision variables.

- **Bounding:** Consists of solving the two LP sub-problems defined above, verifying their results in order to eliminate, or not, the related solutions. A sub-problem is eliminated because either it is infeasible or its value is worse than the current optimum; the sub-problem is retained only in the case of obtaining a feasible solution with a more favorable value than the current optimum, the incumbent value and the associated solution. In the latter situation, if the solution fully satisfies the integer decision variables, it will represent the new optimal solution to the ILP problem. Finally, if possible, return to the branching step; or else, stop the B&B method.

The B&B method ends when it is not possible to find any new LP subproblems that will lead to improvements in the incumbent value for the integer solution, assuming that the incumbent solution exists.

7.2.1 The Furniture Factory Integer Problem

A furniture factory builds tables (t) at a profit of 40 euros per table, and chairs (c) at a profit of 30 euros per chair (Figure 7.1).

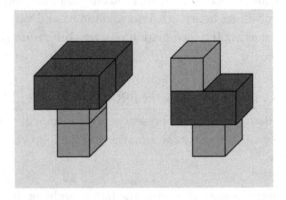

FIGURE 7.1   Tables and chairs for the furniture factory's integer problem.

If the availability of the small-red pieces is 8009 and the availability of the big-blue pieces is 6007, what integer combination of tables and chairs maximizes the total profit?

The ILP problem follows:

$$\max Z = 40t + 30c$$

subject to

$$\begin{cases} 2t + 1c \le 6007 \\ 2t + 2c \le 8009 \\ c \le 3000 \end{cases}$$

$$t, c \ge 0 \ \text{ and } \ \text{integers}$$

ILP Solution by B&B

I) **Initialization:** Set the provisory value for the maximization ILP, $Z^* = 0$, and solve the relaxed LP (*LPr*).

The *LPr* problem is as follows:

$$\max 40t + 30c$$
$$\text{subject to}$$
$$L2)\ 2t + 1c \le 6007$$
$$L3)\ 2t + 2c \le 8009$$
$$L4)\qquad c \le 3000$$

The *LPr* solution can be related to the solution obtained for the LP problem described in the previous chapters.

o In fact, by direct comparison, one additional small-red part is provided in the second restriction, thus the objective function value increases by 10, since the respective dual value is $y_2 = 10$.

o From the matrix approach and the sensitivity analysis, when the RHS parameter b_2 increases by 1, the second variable (c) also increases by 1, while the first variable t decreases by 0.5.

In this way, and remembering the previous solution for the furniture factory problem:

$$Z = 140150$$
$$t = 2003$$
$$c = 2001$$

Thus, the *LPr* solution is

$$Z_{LPr} = 140160$$
$$t = 2002.5$$
$$c = 2002$$

Note that the *LPr* solution can be easily obtained from many computational applications, and the *LPr* value is the upper bound for the integer problem.

- That is, when adding new restrictions aimed at the integer decision variables, the objective function value will successively worsen.

- The optimality check then consists of completing the search for the integer solution that corresponds to the best *LPr* sub-problem, being the *LPr* values treated in descending order.

In addition, variable t takes a non-integer value, $t = 2002.5$, and the branching procedure is implemented in the next step.

II) Until the problem is complete, do:

Branching #1: Two LP sub-problems are generated by adding new restrictions on the non-integer variable t; namely,

$$t \leq 2002 \text{ and } t \geq 2003.$$

Bounding #1: The *LPr* sub-problems #1 and #2 are upper bounded by the related values (Table 7.1).

TABLE 7.1   Branching the Relaxed LP, *LPr*

Sub-Problem #1	Sub-Problem #2
max 40 t + 30 c	max 40 t + 30 c
subject to	subject to
L2) 2 t + 1 $c \leq 6007$	L2) 2 t + 1 $c \leq 6007$
L3) 2 t + 2 $c \leq 8009$	L3) 2 t + 2 $c \leq 8009$
L4) $c \leq 3000$	L4) $c \leq 3000$
L5) $t \leq 2002$	L5) $t \geq 2003$

The solution to *Sub-Problem #1* is

$$Z_1 = 140155$$

$$t = 2002$$

$$c = 2002.5$$

o This solution presents a non-integer value for variable c, while the objective function value Z_1 is a little worse than the *LPr* value; therefore, *Sub-Problem* #1 is still under consideration for additional analysis, i.e., for additional branching in variable c.

For *Sub-Problem* #2, the solution is

$$Z_2 = 140150$$

$$t = 2003$$

$$c = 2001$$

o This solution presents integer values for both variables t and c; then *Sub-Problem* #2 is no longer under consideration for additional analysis because additional branching on the decision variables is not necessary.

o Noting that the objective function value Z_2 is a little worse than the *LPr* value, this value is the new incumbent value, $Z^* = Z_2 = 140150$, since it corresponds to an integer solution, the incumbent solution.

The incumbent value Z^* is also utilized to fathom, skipping additional analysis, all the sub-problems with a value less than the incumbent; in fact, adding new restrictions would reduce the associated values, and improvements to the incumbent would not be possible.

Nevertheless, *Sub-Problem* #1 is still under consideration for additional analysis, because its value, 140155, is greater than the incumbent value; therefore, additional branching for variable c will be addressed in the next step.

Branching #2: For *Sub-Problem* #1, two new LP sub-problems are generated, again introducing new restrictions, but this time related to the non-integer variable c; namely, the restrictions

$$c \leq 2002 \text{ and } c \geq 2003.$$

Bounding #2: The *LPr* sub-problems #1.1 and #1.2 are upper bounded by the related values (Table 7.2).

TABLE 7.2   Branching *Sub-Problem #1*

Sub-Problem #1.1	Sub-Problem #1.2
max $40\,t + 30\,c$	max $40\,t + 30\,c$
subject to	subject to
L2) $2\,t + 1\,c \le 6007$	L2) $2\,t + 1\,c \le 6007$
L3) $2\,t + 2\,c \le 8009$	L3) $2\,t + 2\,c \le 8009$
L4) $c \le 3000$	L4) $c \le 3000$
L5) $t \le 2002$	L5) $t \le 2002$
L6) $c \le 2002$	**L6) $c \ge 2003$**

For *Sub-Problem* #1.1, the solution is

$$Z_{1.1} = 140140$$

$$t = 2002$$

$$c = 2002$$

- This solution also presents integer values for both variables t and c; then *Sub-Problem* #1.1 is fathomed and is no longer under consideration.

- Noting also that the objective function value $Z_{1.1}$ is worse than the incumbent value, $Z^* = Z_2 = 140150$, the incumbent solution remains the same.

For *Sub-Problem* #1.2, the solution is

$$Z_{1.2} = 140150$$

$$t = 2001.5$$

$$c = 2003$$

- This solution presents a non-integer value for the variable t; therefore, *Sub-Problem* #1.2 is still under consideration for additional analysis, i.e., for additional branching in variable t.

In fact, *Sub-Problem* #1.2 is still under consideration because its value, 140150, is equal to the incumbent value; then, additional branching for variable t can result in multiple solutions; this procedure will be addressed in the next step.

Branching #3: For *Sub-Problem* #1.2, two new LP sub-problems are generated, adding a new restriction related to the non-integer variable t, namely,

$$t <= 2001 \text{ and } t \geq 2002.$$

Bounding #3: The *LPr* sub-problems #1.2.1 and #1.2.2 are upper bounded by the related values (Table 7.3).

TABLE 7.3   Branching *Sub-Problem* #1.2

Sub-Problem #1.2.1	*Sub-Problem #1.2.2*
max $40\,t + 30\,c$	max $40\,t + 30\,c$
subject to	subject to
L2) $2\,t + 1\,c \leq 6007$	L2) $2\,t + 1\,c \leq 6007$
L3) $2\,t + 2\,c \leq 8009$	L3) $2\,t + 2\,c \leq 8009$
L4) $c \leq 3000$	L4) $c \leq 3000$
L5) $t \leq 2002$	L5) $t \leq 2002$
L6) $c \geq 2003$	L6) $c \geq 2003$
L7) $t \leq 2001$	**L7) $t \geq 2002$**

For *Sub-Problem* #1.2.1, the solution is

$$Z_{1.2.1} = 140145$$

$$t = 2001$$

$$c = 2003.5$$

- This solution presents a non-integer value for variable c; however, *Sub-Problem* #1.2.1 is fathomed because the objective function value $Z_{1.2.1}$ is worse than the incumbent value, $Z^* = Z_2 = 140150$.

For *Sub-Problem* #1.2.2, a feasible solution could not be found.

  o The production levels are directly constrained in this sub-problem ($t = 2002$, $c = 2003$) and they do not satisfy the second restriction, because such production levels would require 8010 small-red pieces.

  o *Sub-Problem* #1.2.2 is infeasible, and it is fathomed too.

III) **Optimality:**

- o There are no remaining sub-problems to analyze, the search is complete.

- o Then, the B&B method stops!

The current incumbent is optimal, and the optimal integer solution is

$$Z^* = Z_2 = 140150$$

$$t = 2003$$

$$c = 2001$$

The B&B solution tree is also a typical representation of the B&B method. The tree for the furniture factory's integer problem is shown in Figure 7.2.

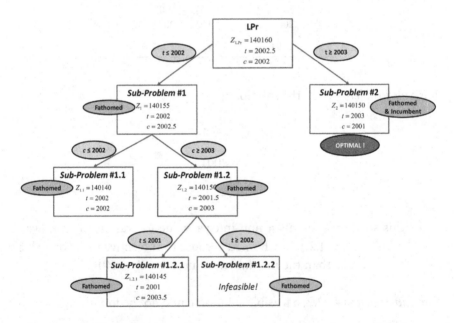

FIGURE 7.2   B&B solution tree for the furniture factory's integer problem.

7.3 MODELING WITH BINARY VARIABLES

In this section, the utilization of binary variables {0, 1} to formulate logic-based restrictions (e.g., logic OR, AND, XOR, contingency decisions), as well as to model fixed charge costs, is described. In addition, given the attributes of binary logic associated with the binary decision variables, these variables can be combined to form more complex logical propositions.

The decision variables y_i present the binary form:

$$y_i = \begin{cases} 1, & \text{positive decision, } \textit{True} \\ 0, & \text{negative decision, } \textit{False} \end{cases}$$

- **Logic OR**

  The logic function OR for two decision variables is shown in Table 7.4, with a *True* result when at least one of the decisions is *True*.

  TABLE 7.4   The Logic Function OR

A	B	OR
T	T	T
T	F	T
F	T	T
F	F	F

  Adding two binary variables, the result is positive when at least one of the binary variables is positive:

  $$y_1 + y_2 \geq 1$$

  When addressing an enlarged set of n decision variables, the result is still positive when at least one of the binary variables is positive:

  $$y_1 + y_2 + \cdots + y_n \geq 1$$

  that is,

  $$\sum_{i=1}^{n} y_i \geq 1$$

- **Logic AND**

  The logic function *AND* for two decision variables is shown in Table 7.5, with a *True* result only when the two decisions are *True*.

TABLE 7.5 The Logic Function *AND*

A	*B*	*AND*
T	T	T
T	F	F
F	T	F
F	F	F

The result is positive only when the two binary variables are positive:

$$y_1, y_2 \geq 1$$

Adding the two positive binary variables, then the sum is

$$y_1 + y_2 \geq 2$$

When addressing an enlarged set of *n* decision variables, the result is positive when all the binary variables are positive:

$$y_1, y_2, \ldots, y_n \geq 1$$

Adding the *n* positive binary variables, the sum thus becomes

$$y_1 + y_2 + \cdots + y_n \geq n$$

that is

$$\sum_{i=1}^{n} y_i \geq n$$

- **Logic *XOR***

The logic function *XOR* for two decision variables is shown in Table 7.6, with a *True* result when the first decision is *True* and the second decision is *False*, and vice versa:

TABLE 7.6 The Logic Function *XOR*

A	*B*	*XOR*
T	T	F
T	F	T
F	T	T
F	F	F

XOR considers two mutually exclusive decisions corresponding to the complementary values for the two binary variables:

$$y_1 = 1 - y_2$$

Noting that the purpose is to select one and only one of the two options, then the related sum is

$$y_1 + y_2 = 1$$

When addressing a mutually exclusive enlarged set of *n* decisions, the purpose is to select one and only one from the *n* options; similarly, the sum is

$$y_1 + y_2 + \cdots + y_n = 1$$

that is

$$\sum_{i=1}^{n} y_i = 1$$

Optionally, it can be required to select a subset of *k* alternatives from all the set with *n* options; for that

$$\sum_{i=1}^{n} y_i = y_1 + y_2 + \cdots + y_n = k$$

Note that the XOR operation can also implement mutually exclusive restrictions, as well as discrete values for equipment selection (e.g., volumes sizing, heat capacities), in this way increasing the modeling flexibility through binary variables.

- **Logic *Equivalence***

The logic function *Equivalence* for two decision variables is shown in Table 7.7, with a *True* result when the two decisions are equivalent; that is, they are both *True* or both *False*:

TABLE 7.7 The Logic Function *Equivalence*

A	B	*Equivalence*
T	T	T
T	F	F
F	T	F
F	F	T

The *Equivalence* result is positive only when the two binary variables are equal:

$$y_1 = y_2$$

By manipulating the variables on the left-hand side, as is usual in LP, then

$$y_1 - y_2 = 0$$

Note that the *Equivalence* operation is complementary to logic *XOR*, and it can also be useful for scheduling operations, allocating operators–machines, sequencing activities and tasks, and optimizing large and complex projects.

- **Fixed charge costs**

  One of the most important applications of binary variables is related to the so-called fixed charge costs. Typically, these costs are incurred regardless of the activity level; for example, all costs associated with cleaning up, setting up, and warming up, as well as maintenance costs can be included in this category. The variable costs or marginal costs are related to direct costs that vary with the produced quantities, for example, raw materials, packaging, and utilities. Fixed charge costs are often utilized due to their straightforward application in real life, in particular to model scale economies. In this situation, the marginal cost decreases when the activity level increases, and scale economies can be represented through a linear formulation that associates binary variables with fixed charge costs.

Given the production level x, and the fixed and variable costs f and c, respectively, the total cost, C_t, can be defined through the following relation:

$$\left.\begin{aligned} C_t &= c.x + f.y \\ x &\le Q.y \,, \quad y \in \{0,1\} \end{aligned}\right\}$$

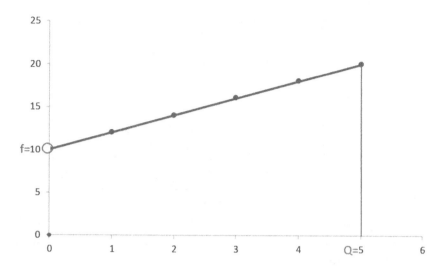

FIGURE 7.3   Total cost (C_t) for a production level with upper-bound ($Q = 5$), fixed charge ($f = 10$), and variable ($c = 2$) components.

For that, y is the binary decision variable that indicates the fixed charge cost associated with any production level greater than 0, and whose maximum level is Q. In graphical terms, assuming $Q = 5$, a fixed charge cost, $f = 10$, and a variable or marginal cost, $c = 2$, then the total cost, C_t, evolution is presented in Figure 7.3.

Note that no production is indicated with a black dot in the axis origin; in fact, the total cost is zero, $C_t = 0$, because the production level is zero, $x = 0$ and the binary variable is zero, $y = 0$. In this case, a fixed charge cost is not incurred.

7.4 SOLVING BINARY INTEGER PROGRAMMING PROBLEMS

The verification or enumeration of all the BIP solutions can be a very time-consuming procedure, or even impossible within real-world needs. The number of binary solutions grows exponentially, 2^n, with the number n of binary variables, as presented in Table 7.8.

As the number of BIP solutions grows exponentially with the number of binary variables, and although many of these solutions will not be examined in the B&B method increasing complexity can be expected. In addition, the number of BIP solutions can be too high to handle, or *overflow* can occur with an everyday computer, as in Table 7.8.

TABLE 7.8   Number of Solutions for a BIP
Problem with n Binary Variables

Binary variables (n)	BIP solutions (2^n)
10	1 024
20	1 048 576
30	1 073 741 824
50	1 125 899 906 842 620
100	1,267650600228230E+30
1000	1,071508607186270E+301
10000	*Overflow*

Although B&B method can assume different particularities, solving a LP problem with binary variables can be summarized in the following pair of fundamental steps:

- **Branching**: Consists of relaxing the binary restrictions and solving the father problem, the relaxed LP; then, create a pair of LP sub-problems by selecting one of the binary variables and fixing its value (e.g., 0, in one case; 1, in the other case). The strategy associated with the selection of the sub-problem and the binary variables to treat depends on the problem at hand.

- **Bounding**: Consists of solving the two LP sub-problems defined previously, then verifying their values in order to eliminate, or not, the related solutions. A sub-problem is eliminated if its solution is infeasible, or if its value is worse than the incumbent value; the sub-problem is retained only if a feasible binary solution with a more favorable value than the incumbent solution is presented. Therefore, if the current solution presents only binary values, it will represent the new optimal solution to the BIP problem, that is, the incumbent value and the associated solution. Finally, if possible, return to the branching step; or else, stop the B&B method.

The method ends when it is not possible to find any new LP sub-problems that will lead to improvements in the value of the binary solution, in the event that such an incumbent solution exists.

7.4.1 BIP Solution by B&B

Steps I and III, respectively, concerning initialization and optimality are similar to those of MILP in Section 7.2, but some differences occur in the iteration cycle in step II. For that, only the second step is presented now, with the alterations highlighted in gray:

II) Until the problem is complete, do:

1) **Branching:** Select the LPr sub-problem with the largest value; generate two sub-problems by fixing the next binary variable in 0 (one sub-problem) and in 1 (the other sub-problem).

2) **Bounding:** The generated LPr sub-problems are upper bounded by the related values.

3) **Fathoming:** An LPr sub-problem is fathomed when:

a) The value is smaller than the incumbent value, $Z_{LPr} \leq Z^*$.

b) It is infeasible.

4) **Updating:** The incumbent solution is updated when the LPr solution is binary and its value is larger than the prior incumbent, $Z^* = Z_{LPr}$.

Re-apply the fathoming step to all the remaining sub-problems.

7.4.2 The Furniture Factory Binary Integer Problem

For illustration purposes, the decision variables associated with tables and chairs are treated as binary variables, 0 or 1; the purpose is to better show the quantitative reduction driven by reducing the solutions space.

The BIP problem follows:

$$\max Z = 40t + 30c$$

subject to

$$\begin{cases} 2t + 1c \leq 6007 \\ 2t + 2c \leq 8009 \\ c \leq 3000 \end{cases}$$

$$t, c \in \{0,1\}$$

7.4.3 The Problem Solution by B&B

I) **Initialization:** Set the provisory value for the maximization BIP, $Z^* = 0$, and solve the relaxed LP (*LPr*).

$$\max 40t + 30c$$

subject to

$$\text{L2) } 2t + 1c \leq 6007$$

$$\text{L3) } 2t + 2c \leq 8009$$

$$\text{L4) } c \leq 3000$$

$$\text{L5) } t \leq 1$$

$$\text{L6) } c \leq 1$$

The *LPr* solution is trivial:

$$Z_{LPr} = 70$$

$$t = 1$$

$$c = 1$$

Note that

- When adding new restrictions aimed at the binary decision variables, their *LPr* values are upper bounded by 1.

- The solution space thus is quite limited, since decision variables can take values from 0 to 1, and the objective function value is greatly reduced.

- A binary solution is obtained, and the problem is fathomed; additional bounds would successively worsen the objective function, then this is the optimal solution.

7.5 CONCLUDING REMARKS

ILP problems are observed in many real situations that are widely disseminated in the literature. In fact, ILP problems frequently appear in real life, formulated as MILP or BIP problems, representing many situations in which either LP proportionality or variables divisibility does not fit.

In addition, the utilization of binary variables {0, 1} to formulate logic-based restrictions (logic *AND*, *OR*, *XOR*; equivalent and contingency decisions), as well as to model fixed charge costs, is presented. The multiple combinations of binary variables and linear restrictions can introduce new dynamics in BIP with wide real-world applications (e.g., capacity expansion, facility location, long-term investment, short-term scheduling).

The furniture factory integer problem is modeled through ILP, by constraining the decision variables to integer values. The B&B method is presented and utilized to treat both the ILP and BIP versions of this problem. It is verified that

- ILP allows qualitative improvement due to more realistic solutions (integer values) and to better modeling capabilities.

- By using the *LPr* value as a bounding value, and successively bounding the solutions space, the optimal value is successively reduced.

- As illustrated in the BIP instance, a severe quantitative reduction results from over-constraining the solutions space, and the BIP solution is trivial because the alternatives at hand are trivial too.

Therefore, the reduction approach in B&B also quantitatively constrains the optimal ILP solution, and it follows the opposite reasoning of a generalization procedure. Notwithstanding that the ILP framework is being quantitatively reduced from the LP approaches, the logic-based modeling features and the integrity attributes largely enhance decision-making. For that, decision makers often use ILP, even for large and complex problems.

The solution to large ILP problems is very difficult, and a number of relevant real-world problems that often occur (e.g., flowshop, jobshop, knapsack, traveling salesman) have been studied in depth. Beyond B&B, several decomposition methods have gained in importance by solving large ILP problems, as they explore certain mathematical structures that occur in such problems. For example, Lagrangean relaxation allows the analysis of a large network of interacting sub-problems; these sub-problems can be autonomously treated due to their smaller size, and their solutions lead to the efficient solution of the original ILP problem.

Regarding decision-making, ILP assumes that one single decision maker selects the optimal solution by optimizing the objective function value. Once again, and similarly to LP, one alternative is being selected by considering one single attribute (e.g., maximizing gain; minimizing cost). By assuming the existence of a second decision maker, which autonomously selects pertinent alternatives and impacts system results, Game Theory thus is a very important topic for decision support and will be presented in Chapter 8.

Game Theory

IN THIS CHAPTER, A brief discussion on game theory is provided, broadening the decision framework by addressing two decision-makers: the two players as they are usually known. Beyond the key ideas of the two-player constant-sum game and zero-sum game, the associated linear programming (LP) formulations are developed. The duality approach associated with such LP complementary versions is also addressed, including both Max-Min and Min-Max versions. Concerning the direct support for decision-making, the treatment of both mixed strategies and dominant strategies is an important tool that is also presented in this chapter.

8.1 INTRODUCTION

A game occurs when more than one decision-maker simultaneously participates in a decision-making process, and because of that game theory is perceived as a generalization of the decision framework with one single decision-maker. Typically, it is observed that:

- Each decision-maker can select different alternatives, that is, each player can take different options from a pre-defined set; in this way, the space of alternatives for each decision-maker is well defined and the players have the same information.

- The result obtained in each move depends on the choices made by all of the players. Each decision-maker selects an alternative at the same

DOI: 10.1201/9781315200323-8

time as the other, and the outcome of the game depends on such simultaneous selection.

- Each decision-maker selects their own alternative without any knowledge of the choices made by the other players.

Common applications of game theory include, among many others:

- Bargaining: aimed at obtaining a more favorable solution for one of the players; for example, a simple game in which one person buys something that another sells, or an industry company renting out a production line to a distribution company with its own "white brands".

- Conflict reduction: intended to reduce the risk of serious levels of conflict (or even violence) that arise when certain options are taken; for example, a game in which there is an outside intervener, such as a government body with coordination and regulation functions, which aims at reducing the social conflict between worker unions and enterprise associations.

- Cooperation: designed to support all players in a way that their options correspond to a solution that maximizes the overall net benefit; for example, a game in which pacts are established between players, as in the coalition between civil construction companies to address large and complex projects of highways or airports; or even the international cooperation of countries and states to access medicines, vaccines, or emergency rescue in face of a cataclysm.

8.2 CONSTANT-SUM GAME

Constant-sum games assume the total amount distributed by the two players is constant, that is, the amount increased by one player is diminished by the other, the amount the first player gains the second loses. Due to the conflicting nature a constant-sum game implicitly assumes, no cooperation between players is expected. These games are also considered a type of non-zero sum game that is reduced to a specific and constant amount.

The normal form of a constant-sum game assumes that the game has a set of options (i, j), for *Player-One* and *Player-Two*, respectively, as well as the results received for the related set, U_{ij}.

A pessimistic behavior is also assumed for both players; they both know that the choices made by the other player will leave them in an unfavorable position. In a simple way, it is important that both players avoid the worst situation; *Player-One* selects the best payment from the worst results that are associated with the alternatives of *Player-Two*; and vice-versa, *Player-Two* selects the best payment from the worst results that are associated with the alternatives of *Player-One*.

For that, let it be: i) α is the maximum value for the i alternatives of *Player-One*, to be selected from the minimum gain related to the j options of *Player-Two*, and ii) similarly, β is the maximum value for the j options of *Player-Two*, to be selected from the minimum gain related to the i alternatives of *Player-One*.

$$\begin{cases} \alpha = \max_{i} \left(\min_{j} u_{ij} \right) \\ \beta = \max_{j} \left(\min_{i} u_{ij} \right) \end{cases}$$

Looking for a stable situation, the equilibrium point will correspond to the simultaneous choices of the two players when, assuming all the existing alternatives for the second player, it will not be possible for the first player to improve his result; that is, it is not possible to select another option that will lead to a more advantageous result for the first player than the existing one, and vice-versa. Thus, the necessary condition for a stable balance, also known as "saddle stitch", is that the sum of payments for the two players reaches the constant amount, K:

$$\alpha + \beta = K$$

The Nash equilibrium point is a generalization of the "saddle stitch" concept in non-zero sum games. It corresponds to the global set made by the n players' decisions $\{d_1, d_2, \ldots, d_n\}$ in such a way that it is not possible for any player i to select another alternative that will lead to a more advantageous result than the decision d_i, assuming the remaining decisions $\{d_1, d_2, ... \, d_{i-1},$ $d_{i+1}, ..., d_n\}$ for the other $(n-1)$ players.

8.2.1 Constant-Sum Game with Stable Equilibrium

Let us consider a constant-sum game in the normal form, with the game considering a set of alternatives both for *Player-One* (A1–A3) and *Player-Two* (B1–B4); the sum of results the two players receive is constant and equal to 100 (e.g., market quota or sales percentage), as presented in Table 8.1.

TABLE 8.1   Payments Matrix for Constant-Sum Game ($K = 100$)

Payments Matrix	B1	B2	B3	B4
A1	(65,35)	(55,45)	(75,25)	(60,40)
A2	(75,25)	(65,35)	(70,30)	(100,0)
A3	(70,30)	(50,50)	(55,45)	(70,30)

A1, B1: increasing demand, by reducing the sales prices for the furniture parts.

A2, B2: *ditto*, by increasing promotion actions, marketing, and communication.

A3, B3: *ditto*, through a suitable combination of prices and promotion actions.

B4: *ditto*, introducing gifts, vouchers for a new design line of furniture.

Note the payments sum is constant, $K = 100$, the amount increased by *Player-One* is diminished in *Player-Two*; that is, when *Player-One* gains 10, then the *Player-Two* loses 10, and vice-versa. Additionally:

- The space of alternatives for each decision-maker is defined, and the two players have the same information.

- *Player-One* selects his own alternative without knowing the option made by *Player-Two*, and vice-versa.

- The result obtained in each move depends on the options the two players simultaneously made.

Player-One: The Minimum Value per Line

A pessimistic profile is assumed by *Player-One*, by supposing the results will be the most unfavorable for each of the alternatives under consideration. It is thus important that *Player-One* avoids the worst situation; therefore, *Player-One* selects the best payment from the worst results that originated from the choices made by *Player-Two*.

TABLE 8.2   Constant-Sum Game: Payments Matrix for *Player-One*

Payments *Player-One*	B1	B2	B3	B4
A1	65	**55**	75	60
A2	75	**65**	70	100
A3	70	**50**	55	70

From *Player-One*'s point of view, and for each of the alternatives A1–A3 at hand, the worst results would occur if *Player-Two* selected alternative B2, those results are bolded in Table 8.2. In the first step, the pessimistic behavior corresponds to the search for a minimum value at each line of the payments matrix. For the second step, *Player-One* selects the maximum value from these minima; that is, the best one from the worst results, as presented in Table 8.3.

TABLE 8.3   Constant-Sum Game: Max-Min Approach for *Player-One*

Payments *Player-One*	Pessimistic (Max-Min)
A1	55
A2	**65**
A3	50

Within this Max-Min approach in two steps, *Player-One* selects the alternative A2, with a gain of 65 ($\alpha = 65$).

Player-Two: The Minimum Value per Column

A pessimistic profile is also assumed by *Player-Two*, by supposing the results will be the most unfavorable. It is also important for *Player-Two* to avoid the worst situation by selecting the best payment from the worst results that originated from the choices made by *Player-One*.

In the first step, the pessimistic behavior of *Player-Two* corresponds to finding the minimum value at each column of the payments matrix for each of the alternatives B1–B4. The worst results associated with *Player-One*'s choices are bolded in Table 8.4.

TABLE 8.4   Constant-Sum Game:
Payments Matrix for *Player-Two*

Payments *Player-Two*	B1	B2	B3	B4
A1	35	45	**25**	40
A2	**25**	**35**	30	**0**
A3	30	50	45	30

For the second step, *Player-Two* selects the maximum value from these minima; that is, the best of the worst results, as presented in Table 8.5.

TABLE 8.5   Constant-Sum Game: Max-Min
Approach for *Player-Two*

Payments *Player-Two*	B1	B2	B3	B4
Pessimistic (Max-Min)	25	**35**	25	0

Within this Max-Min approach, *Player-Two* is selecting the alternative B2, gaining 35 ($\beta = 35$).

The equilibrium point is achieved, the sum of the best options for *Player-One* ($\alpha = 65$) and *Player-Two* ($\beta = 35$) reach the constant-sum, $K = 100$. The situation is stable because neither *Player-Two* nor *Player-One* can improve their gains by selecting other options, and thus they both maintain the options pair (A2, B2) with the associated result. The rationale follows:

- In fact, *Player-One* takes option A2 and gains 65, because this is the best option when assuming *Player-Two* would choose option B2; otherwise, the gain would be reduced to 55 or 50.

- Similarly, *Player-Two* takes option B2, gaining 35, because this is the best option when assuming *Player-One* would choose option A2; otherwise, the gain would be reduced to 25, 20, or 0.

In summary, with options A2 for *Player-One* and B2 for *Player-Two*, the result is stable and equilibrium is reached; these options are bolded in Table 8.6.

TABLE 8.6   Constant-Sum Game: Stable Equilibrium

Payments Matrix	B1	B2	B3	B4
A1	(65,35)	(55,45)	(75,25)	(60,40)
A2	(75,25)	**(65,35)**	(70,30)	(100,0)
A3	(70,30)	(50,50)	(55,45)	(70,30)

8.2.2 Constant-Sum Game with Cycle

Let us consider a payments matrix with a single alteration, associated with the result of the options pair (A2, B3); the result is now (60,40). The sum is still constant and equal to 100. This single alteration is marked in bold in Table 8.7.

TABLE 8.7   Altered Matrix for Constant-Sum Game with Cycle

Payments Matrix	B1	B2	B3	B4
A1	(65,35)	(55,45)	(75,25)	(60,40)
A2	(75,25)	(65,35)	**(60,40)**	(100,0)
A3	(70,30)	(50,50)	(55,45)	(70,30)

Similarly, through a pessimistic approach in two steps for *Player-One*: i) find the minimum value at each line, and ii) select the maximum of those minima (Max-Min). The final outcome for *Player-One* diminishes to $\alpha = 60$; the outcome suffers from the impact of a data alteration since in the prior instance it was 65. The related results are presented, respectively, in Table 8.8*a* and Table 8.8*b*.

TABLE 8.8a,b   Constant-Sum Game with Cycle: Max-Min Approach for *Player-One*

Payments *Player-One*	B1	B2	B3	B4
A1	65	55	75	60
A2	75	65	60	100
A3	70	50	55	70

Payments *Player-One*	Pessimistic (Max-Min)
A1	55
A2	**60**
A3	50

Player-Two, also using the pessimistic approach, searches for the minimum value at each column and then selects the maximum of the minima (Max-Min). The related results are presented, respectively, in Table 8.9a, and Table 8.9b. The final outcome for *Player-Two* is still the same as in the prior instance, $\beta = 35$.

TABLE 8.9a,b Constant-Sum Game with Cycle:
Max-Min Approach for *Player-One* (Search by Column)

Payments *Player-Two*	B1	B2	B3	B4
A1	35	45	25	40
A2	25	35	40	0
A3	30	50	45	30
Payments *Player-Two*	**B1**	**B2**	**B3**	**B4**
Pessimistic (Max-Min)	25	35	25	0

With the same selected options as in the first instance, alternative A2 for *Player-One* and alternative B2 for *Player-Two*, the result at now still is the same for the previous instance, as presented in the altered payments matrix in Table 8.10.

TABLE 8.10 Altered Matrix for Constant-Sum Game with Cycle: Not Stable, First Move

Payments Matrix	B1	B2	B3	B4
A1	(65,35)	(55,45)	(75,25)	(60,40)
A2	(75,25)	(65,35)	(60,40)	(100,0)
A3	(70,30)	(50,50)	(55,45)	(70,30)

However, equilibrium is not achieved, because the sum of the best options for *Player-One* ($\alpha = 60$) and *Player-Two* ($\beta = 35$) does not reach the reference value, 100.

The instance is not stable because *Player-Two* can improve his gain by selecting option B3, while it can be assumed *Player-One* maintains the best option A2. The rationale follows:

- *Player-One* remains at option A2 and gains 65, because this is the best option when assuming *Player-Two* would maintain option B2.

- However, *Player-Two* does not continue with option B2. It is more advantageous to switch to option B3 and gain 40 when assuming *Player-One* would maintain option A2; otherwise, the gain would only be 35.

The second move follows with a gain of 60 for *Player-One* and 40 for *Player-Two*, associated with the options A2 and B3, as presented in Table 8.11.

TABLE 8.11   Altered Matrix for Constant-Sum Game with Cycle: Second Move

Payments Matrix	B1	B2	B3	B4
A1	(65,35)	(55,45)	(75,25)	(60,40)
A2	(75,25)	**(65,35)**	**(60,40)**	(100,0)
A3	(70,30)	(50,50)	(55,45)	(70,30)

Again, equilibrium is not reached, because the instance is not stable. Now, *Player-One* can improve his gain by selecting option A1, while it can be assumed *Player-Two* would maintain the best option B3. The rationale now:

- *Player-Two* remains at option B3, because this is the best option when assuming *Player-One* would maintain option A2; in this way, *Player-Two* would continue to gain 40.

- But *Player-One* does not continue with option A2 when assuming *Player-Two* would maintain option B3. It is more advantageous to switch to option A1 and gain 75; otherwise, the gain would only be 60.

In the third move, *Player-One* gains 75 and *Player-Two* gains 25, associated with the options A1 and B3, respectively, as shown in the payments matrix in Table 8.12.

TABLE 8.12   Altered Matrix for Constant-Sum Game with Cycle: Third Move

Payments Matrix	B1	B2	B3	B4
A1	(65,35)	(55,45)	**(75,25)**	(60,40)
A2	(75,25)	**(65,35)**	**(60,40)**	(100,0)
A3	(70,30)	(50,50)	(55,45)	(70,30)

Equilibrium is not reached once again, and the instance is still not stable. At the current state, *Player-Two* can improve his gain by selecting option B2, while it can be assumed *Player-One* maintains the best option A1. Note that the players' options are not known *a priori*, and:

- *Player-One* assumes *Player-Two* would maintain option B3 and then the best option is to still remain with option A1; in this way, *Player-One* would continue to gain 75.

- However, *Player-Two* does not continue with option B3 when assuming *Player-One* would maintain option A1. It is more advantageous for *Player-Two* to switch to option B2, and gain 45; otherwise, the gain would only be 25.

The fourth move, representing options A1 and B2 with gains of 55 for *Player-One* and 45 for *Player-Two*, follows in Table 8.13.

TABLE 8.13   Altered Matrix for Constant-Sum Game with Cycle: Fourth Move

Payments Matrix	B1	B2	B3	B4
A1	(65,35)	**(55,45)**	**(75,25)**	(60,40)
A2	(75,25)	**(65,35)**	**(60,40)**	(100,0)
A3	(70,30)	(50,50)	(55,45)	(70,30)

At the fifth move, a cycle of dynamic interactions is observed. *Player-One* can improve his gain by selecting option A2, while it can be assumed *Player-Two* maintains the best option B2. Since the players' options are not known *a priori*, the reasoning follows:

- *Player-Two* assumes *Player-One* would maintain option A1 and then the best option is to remain with option B2; in this way, *Player-Two* would continue to gain 45.

- Nevertheless, *Player-One* does not continue with option A1 when assuming *Player-Two* would maintain option B2. It is more advantageous to alter to option A2 and gain 65; otherwise, the gain would only be 55.

With option A2 for *Player-One* and B2 for *Player-Two*, the fifth state of play is equal to the first state, with gains 65 and 35, respectively. The payments matrix in Table 8.14 represents the initial step for a second cycle of dynamic interactions between the two players.

TABLE 8.14   Altered Matrix for Constant-Sum Game with Cycle: Fifth Move

Payments Matrix	B1	B2	B3	B4
A1	(65,35)	**(55,45)**	**(75,25)**	(60,40)
A2	(75,25)	**(65,35)**	**(60,40)**	(100,0)
A3	(70,30)	(50,50)	(55,45)	(70,30)

In conclusion, the sum of the best options for *Player-One* ($\alpha = 60$) and *Player-Two* ($\beta = 35$) does not reach the constant value, $K = 100$. In this instance, equilibrium cannot be achieved and cycles of dynamic interactions are then observed. This phenomenon can be better explained by transforming the current game into a zero-sum game, by subtracting 50 from the payments of both players.

8.3 ZERO-SUM GAME

Zero-sum games assume the total amount distributed by the two players is not only constant, but also equal to zero. Once again, the amount increased by the first player is subtracted by the other, the amount *Player-One* gains is equal to the amount *Player-Two* loses. Therefore, no cooperation between players is expected, and the second player will only be interested in participating if compensation outside the results matrix is provided.

As with the constant-sum game described in the previous section, the normal form and pessimistic behavior are both assumed in a zero-sum game.

Zero-sum games are also considered a specific type of constant-sum Game, by reducing the constant value K to zero, $K = 0$. The necessary condition for "saddle stitch" and stable balance is also updated to zero:

$$\alpha + \beta = 0$$

8.3.1 Zero-Sum Game with Stable Equilibrium

For a stable equilibrium game, the payments matrix for the zero-sum game is obtained by subtracting 50 from the payments of the two players. The constant-sum game is then reduced to a zero-sum game. The matrix elements can be seen as the alterations from a reference value, *e.g.*, 50. For example, suppose a market's quota or the sales percentage amount is equally divided between the players, then each player has 50 of the total value, 100. The alterations are related to the reference value: for example, if *Player-One* gains 5, *Player-Two* loses 5, the pair (55,45) is reduced to (5,–5); if *Player-One* gains 10, *Player-Two* loses 10, the pair (60,40) is reduced to (10,–10); and so on, as presented in the Table 8.15.

TABLE 8.15 Payments Matrix for Zero-Sum Game ($K = 0$)

Payments Matrix	B1	B2	B3	B4
A1	(15,–13)	(5,–5)	(25,–25)	(10,–10)
A2	(25,–25)	(15,–15)	(20,–20)	(50,–50)
A3	(20,–20)	(0,–0)	(5,–5)	(20,–20)

Player-One: The Minimum Value per Line

Once again, *Player-One* is assuming a pessimistic behavior and avoiding the worst situation. Therefore, in the first step *Player-One* searches for the minimum value at each line of the payments matrix; those results are bolded in Table 8.16.

TABLE 8.16 Zero-Sum Game: Payments Matrix for *Player-One*

Payments *Player-One*	B1	B2	B3	B4
A1	15	**5**	25	10
A2	25	**15**	20	50
A3	20	**0**	5	20

For the second step, *Player-One* selects the maximum value from these minima, as presented in Table 8.17.

TABLE 8.17 Zero-Sum Game: Max-Min Approach for *Player-One*

Payments *Player-One*	Pessimistic (Max-Min)
A1	5
A2	**15**
A3	0

With this approach, *Player-One* is selecting the alternative A2, with a gain of 15 ($\alpha = 15$).

Player-Two: The Minimum Value per Column
A pessimistic profile is also assumed by *Player-Two*. The first step corresponds to the minimum value in each column of the payments matrix. The worst results associated with *Player-One*'s choices are bolded in Table 8.18.

TABLE 8.18 Zero-Sum Game: Payments Matrix for *Player-Two*

Payments Player-2	B1	B2	B3	B4
A1	−15	−5	**−25**	−10
A2	**−25**	−15	−20	**−50**
A3	−20	0	−5	−20

For the second step, *Player-Two* selects the maximum value from these minima; that is, the best of the worst results, as presented in Table 8.19.

TABLE 8.19 Zero-Sum Game: Max-Min Approach for *Player-Two*

Payments Player-Two	B1	B2	B3	B4
Pessimistic (Max-Min)	−25	−15	−25	−50

Within this Max-Min approach, *Player-Two* is selecting the alternative B2, losing 15 ($\beta = -15$).

With option A2 for *Player-One* and B2 for *Player-Two*, the result is stable and equilibrium is achieved, as presented in Table 8.20.

TABLE 8.20 Zero-Sum Game: Stable Equilibrium

Payments Matrix	B1	B2	B3	B4
A1	(15,–15)	(5,–5)	(25,–25)	(10,–10)
A2	(25,–25)	**(15,–15)**	(20,–20)	(50,–50)
A3	(20,–20)	(0,–0)	(5,–5)	(20,–20)

The equilibrium point is reached, note the sum of the best options for *Player-One* ($\alpha = 15$) and *Player-Two* ($\beta = -15$) is the zero-sum, $K = 0$. The situation is stable because neither *Player-Two* nor *Player-One* can improve their gains by selecting other options, and thus they both maintain the options pair (A2, B2) with the associated result. The rationale follows:

- In fact, *Player-One* chooses option A2 and gains 15, because this is the best option when assuming *Player-Two* would maintain option B2; otherwise, the gain would be reduced to 5 or 0.

- Similarly, *Player-Two* chooses option B2, only losing –15, because this is the best option when assuming *Player-One* would choose option A2; otherwise, the gain (that is, the loss of market quota, or the amount of lost sales) would be reduced to –25, –20, or –50.

8.3.2 Zero-Sum Game with Cycle

Again, let us consider the payments matrix with a single alteration, associated with the result of the options pair (A2, B3); the result (60,40) is now transformed to (10,–10) and the associated sum is still constant and equal to 0. This alteration is bolded in Table 8.21.

TABLE 8.21 Altered Matrix for Zero-Sum Game with Cycle

Payments Matrix	B1	B2	B3	B4
A1	(15,–15)	(5,–5)	(25,–25)	(10,–10)
A2	(25,–25)	(15,–15)	**(10,–10)**	(50,–30)
A3	(20,–20)	(0,0)	(5,–5)	(20,–20)

With the pessimistic approach in two steps for *Player-One*, the first step is to find the minimum value at each line and then select the maximum of these minima (Max-Min). The related results are bolded in Table 8.22.

TABLE 8.22 Zero-Sum Game with Cycle:
Max-Min Approach for *Player-One*

Payments Player-One	B1	B2	B3	B4
A1	15	5	25	10
A2	25	15	**10**	50
A3	20	0	5	20

Player-Two also uses the pessimistic approach to find the minimum value in each column and then selects the maximum of these minima (Max-Min). The two-step results are shown in Table 8.23.

TABLE 8.23 Zero-Sum Game with Cycle: Max-
Min Approach for *Player-Two* (Search by Column)

Payments Player-Two	B1	B2	B3	B4
A1	–15	–5	–25	–10
A2	–25	**–15**	–10	–50
A3	–20	0	–5	–20

Combining option A2 for *Player-One* and B2 for *Player-Two*, the result is the same as in the previous instance, as presented in Table 8.24.

TABLE 8.24 Altered Matrix for Zero-Sum Game with Cycle: Not Stable, First Move

Payments Matrix	B1	B2	B3	B4
A1	(15,–15)	(5,–5)	(25,–25)	(10,–10)
A2	(25,–25)	**(15,–15)**	(10,–10)	(50,–50)
A3	(20,–20)	(0,–0)	(5,–5)	(20,–20)

However, equilibrium is not reached, the sum of the best options for *Player-One* ($\alpha = 10$) and *Player-Two* ($\beta = -15$) does not reach a zero-sum, the sum is $K = -5$.

The instance is not stable because *Player-Two* can improve his gain by selecting option B3, while it can be assumed *Player-One* maintains the best option A2. The rationale follows:

- *Player-One* remains at option A2 and gains 15, because this is the best option when assuming *Player-Two* would maintain option B2.

- However, *Player-Two* does not continue with option B2. It is more advantageous to switch to option B3 when assuming *Player-One* would maintain option A2, and improve the loss to –10; otherwise, the loss would be –15.

The second move follows with a gain of 10 for *Player-One* and a loss of –10 for *Player-Two*, associated with the options A2 and B3, respectively, as presented in Table 8.25.

TABLE 8.25   Altered Matrix for Zero-Sum Game with Cycle: Second Move

Payments Matrix	B1	B2	B3	B4
A1	(15,–15)	(5,–5)	(25,–25)	(10,–10)
A2	(25,–25)	**(15,–15)**	**(10,–10)**	(50,–50)
A3	(20,–20)	(0,–0)	(5,–5)	(20,–20)

Again, equilibrium is not reached, because the instance is not stable. Now, *Player-One* can improve his gain by selecting option A1, while it can be assumed *Player-Two* maintains the best option B3. The rationale now:

- *Player-Two* remains at option B3, because this is the best option when assuming *Player-One* would maintain option A2. In this way, *Player-Two* would continue with a loss of –10.

- But *Player-One* does not continue with option A2 when assuming *Player-Two* would maintain option B3. It is more advantageous to switch to option A1 and gain 25; otherwise, the gain would only be 10.

In the third move, *Player-One* gains 25 and *Player-Two* has a loss of –25, associated with options A1 and B3, respectively. The payments matrix is presented in Table 8.26.

TABLE 8.26   Altered Matrix for Zero-Sum Game with Cycle: Third Move

Payments Matrix	B1	B2	B3	B4
A1	(15,–15)	(5,–5)	**(25,–25)**	(10,–10)
A2	(25,–25)	**(15,–15)**	**(10,–10)**	(50,–50)
A3	(20,–20)	(0,–0)	(5,–5)	(20,–20)

Equilibrium is not reached, and once again the instance is not stable. At the current state, *Player-Two* can improve his gain by selecting option B2, while it can be assumed *Player-One* would maintain the best option A1. The players choose their options simultaneously and they are not known *a priori*. Thus:

- *Player-One* assumes *Player-Two* would maintain option B3 and then the best option is to continue with option A1; in this way, *Player-One* would continue to gain 25.

- However, *Player-Two* does not continue with option B3 when assuming *Player-One* would maintain option A1. It is more advantageous for *Player-Two* to switch to option B2, and in this manner reduce the loss from –25 to –5.

The fourth move, representing options A1 and B2 with a gain of 5 for *Player-One* and a loss of –5 for *Player-Two*, is shown in Table 8.27.

TABLE 8.27　Altered Matrix for Zero-Sum Game with Cycle: Fourth Move

Payments Matrix	B1	B2	B3	B4
A1	(15,–15)	**(5,–5)**	**(25,–25)**	(10,–10)
A2	(25,–25)	**(15,–15)**	**(10,–10)**	(50,–50)
A3	(20,–20)	(0,–0)	(5,–5)	(20,–20)

At the fifth move, a cycle of dynamic interactions is observed again. At this fifth state, *Player-One* can improve his gain by selecting option A2, while it can be assumed *Player-Two* maintains the best option B2. Since the players choose their options simultaneously, and these options are not known *a priori*, the reasoning follows that:

- *Player-Two* assumes *Player-One* would maintain option A1 and then the best option is to still choose option B2; in this way, *Player-Two* would continue to lose only –5.

- Nevertheless, *Player-One* does not continue with option A1 when assuming *Player-Two* would maintain option B2. It is more advantageous to switch to option A2 and improving the gain from 5 to 15.

With options A2 *Player-One* and B2 *Player-Two*, the fifth move is equal to the first state, with symmetric gains of 15 and –15. Table 8.28 represents the initial step for a second cycle of dynamic interactions between the two players.

TABLE 8.28　Altered Matrix for Zero-Sum Game with Cycle: Fifth Move

Payments Matrix	B1	B2	B3	B4
A1	(15,15)	**(5,–5)**	(25,–25)	(10,–10)
A2	(25,–25)	**(15,–15)**	**(10,–10)**	(50,–50)
A3	(20,–20)	(0,–0)	(5,–5)	(20,–20)

In conclusion, the sum of the best options for *Player-One* ($\alpha = 10$) and *Player-Two* ($\beta = -15$) does not reach zero, in fact $K = -5$. Then, equilibrium cannot be achieved, and cycles of dynamic interactions occur.

8.4 MIXED STRATEGIES: LP APPROACH

Note that each player should avoid successively repeating the most preferred option or replicating the same strategies over and over, so that the other players can not foresee what the next move will be.

An important approach is to consider mixing diverse strategies, for example, either randomly or with certain reasoning. Therefore, a player can select a strategies subset by targeting the maximum expected value within the alternatives under study.

Within a two-player constant-sum game, the maximization of *Player-One*'s gains is complemented with the minimization of *Player-Two*'s losses, this is called the primal-dual approach.

8.4.1 Player-One: Mixed Strategies and LP

Addressing the mixed strategies for *Player-One*, a LP model is presented with the objective of maximizing the expected value for the payments, while associating frequencies or probabilities with each one of the alternatives at hand.

Objective
To maximize the expected value, X, for the payments to be received by *Player-One*.

Decision Variables
The decision variables are the probabilities set (p_1, p_2, p_3) that are associated with the options (A1, A2, A3) that *Player-One* can choose.

$$\max\ Z = X + 0p_1 + 0p_2 + 0p_3$$

subject to

$$\begin{aligned}
\text{P)} \quad & p_1 + p_2 + p_3 = 1 \\
\text{B1)} \quad & X \le 15p_1 + 25p_2 + 20p_3 \\
\text{B2)} \quad & X \le 5p_1 + 15p_2 + 0p_3 \\
\text{B3)} \quad & X \le 25p_1 + 20p_2 + 5p_3 \\
\text{B4)} \quad & X \le 10p_1 + 50p_2 + 20p_3
\end{aligned}$$

As required in LP, transposing the variables on the right-hand side to the first member with the contrary (minus) sign, and keep constant parameters for the second member (1 and 0, for this instance):

$$\max Z = X + 0p_1 + 0p_2 + 0p_3$$

subject to

$$\text{P)} \qquad p_1 + \quad p_2 + \quad p_3 = 1$$
$$\text{B1)} \quad X - 15p_1 - 25p_2 - 20p_3 \leq 0$$
$$\text{B2)} \quad X - 5p_1 - 15p_2 \qquad \leq 0$$
$$\text{B3)} \quad X - 25p_1 - 20p_2 - 5p_3 \leq 0$$
$$\text{B4)} \quad X - 10p_1 - 50p_2 - 20p_3 \leq 0$$

The optimal expected value is $Z^* = 15$, and the optimal solution is

$$\begin{bmatrix} X \\ p_1 \\ p_2 \\ p_3 \end{bmatrix}^* = \begin{bmatrix} 15 \\ 0 \\ 1 \\ 0 \end{bmatrix}$$

These probabilistic results are consistent with the stable equilibrium occurring for the first zero-sum where *Player-One* selects alternative A2. Alternative A2 continues to be used because the other alternatives do not add new improvements, they are not more advantageous.

For the cyclic instance (replacing 20 with 10, in the B3-line for probability variable p_2), with dynamic interactions between the two players due to successive improvements and losses for each player, the probabilistic approach changes to $Z^* = 13$ and

$$\begin{bmatrix} X \\ p_1 \\ p_2 \\ p_3 \end{bmatrix}^* = \begin{bmatrix} 13 \\ 0.2 \\ 0.8 \\ 0 \end{bmatrix}$$

These probabilistic results are also consistent with the cyclic interaction that occurs with the altered zero-sum instance; first, because *Player-One* only selects alternatives A1 and A2. In addition, for a large number of moves, these results indicate that *Player-One* selects alternative A1 20% of the time, while selecting A2 in the other 80% of the time.

8.4.2 Player-Two: Mixed Strategies and LP

Addressing the mixed strategies for *Player-Two*, an LP model is presented with the objective of minimizing the expected value for the lost amounts, while associating frequencies or probabilities with each one of the alternatives.

Objective

To minimize the expected value Y for the losses to be paid by *Player-Two*.

Decision Variables

The decision variables are the probabilities set (q_1, q_2, q_3, q_4) that are associated with the options (B1, B2, B3, B4) *Player-Two* can choose.

$$\min W = Y + 0q_1 + 0q_2 + 0q_3 + 0q_4$$

subject to

Q) $q_1 + q_2 + q_3 + q_4 = 1$

A1) $Y \geq 15q_1 + 5q_2 + 25q_3 + 10q_4$

A2) $Y \geq 25q_1 + 15q_2 + 20q_3 + 50q_4$

A3) $Y \geq 20q_1 \qquad + 5q_3 + 20q_4$

And again, as required in LP, manipulate the variables on the right-hand side to the first member with a minus sign:

$$\min W = Y + 0q_1 + 0q_2 + 0q_3 + 0q_4$$

subject to

Q) $q_1 + q_2 + q_3 + q_4 = 1$

A1) $Y - 15q_1 - 5q_2 - 25q_3 - 10q_4 \geq 0$

A2) $Y - 25q_1 - 15q_2 - 20q_3 - 50q_4 \geq 0$

A3) $Y - 20q_1 \qquad - 5q_3 - 20q_4 \geq 0$

The optimal expected value is $W^* = 15$, and the optimal solution is

$$\begin{bmatrix} Y \\ q_1 \\ q_2 \\ q_3 \\ q_4 \end{bmatrix}^* = \begin{bmatrix} 15 \\ 0 \\ 1 \\ 0 \\ 0 \end{bmatrix}$$

These probabilistic results are consistent with the stable equilibrium occurring for the first zero-sum instance; in fact, *Player-Two* selects alternative B2, and then maintains B2 because the other alternatives do not add new improvements.

For the cyclic instance (replacing –20 with –10, in the A2-line for the probability variable q_3), dynamic interactions between the two players occur due to successive improvements and losses for each player. The probabilistic approach changes to $W^* = 13$ and

$$
\begin{bmatrix} Y \\ q_1 \\ q_2 \\ q_3 \\ q_4 \end{bmatrix}^* = \begin{bmatrix} 13 \\ 0 \\ 0.6 \\ 0.4 \\ 0 \end{bmatrix}
$$

These probabilistic results are also consistent with the cyclic interaction that occurs with the altered zero-sum instance; first, because *Player-Two* only selects alternatives B2 and B3. In addition, for a large number of moves, these results can indicate that *Player-Two* selects alternative B2 60% of the time, and B3 for the other 40% of the time.

8.4.3 Primal-Dual Approaches for Zero-Sum Game with Cycle

The maximization LP for *Player-One* for a zero-sum game with cycle is addressed, and the altered instance for the related primal LP in algebraic form is presented:

$$
\max Z = X + 0p_1 + 0p_2 + 0p_3
$$

subject to

P) $\quad\quad\quad p_1 + \quad p_2 + \quad p_3 = 1$

B1) $\quad X - 15p_1 - 25p_2 - 20p_3 \leq 0$

B2) $\quad X - 5p_1 - 15p_2 \quad\quad\quad \leq 0$

B3) $\quad X - 25p_1 - 10p_2 - 5p_3 \leq 0$

B4) $\quad X - 10p_1 - 50p_2 - 20p_3 \leq 0$

As described in Chapter 4, the primal-dual transformation uses the matrix form for the primal LP:

$$\max [Z] = \begin{bmatrix} 1 & 0 & 0 & 0 \end{bmatrix} . \begin{bmatrix} X \\ p_1 \\ p_2 \\ p_3 \end{bmatrix}$$

subject to

$$\begin{bmatrix} 0 & 1 & 1 & 1 \\ 1 & -15 & -25 & -20 \\ 1 & -5 & -15 & 0 \\ 1 & -25 & -10 & -5 \\ 1 & -10 & -50 & -20 \end{bmatrix} . \begin{bmatrix} X \\ p_1 \\ p_2 \\ p_3 \end{bmatrix} \leq \begin{bmatrix} 1 \\ 0 \\ 0 \\ 0 \\ 0 \end{bmatrix}$$

$$\begin{bmatrix} X & p_1 & p_2 & p_3 \end{bmatrix}^T \geq 0$$

Synoptically, the primal-dual transformation for the maximization LP of *Player-One* considers the following steps:

- The primal LP is a maximization problem with "less than or equal to" ($\leq$) restrictions; and the associated dual is a minimization problem with "greater than or equal to" ($\geq$) restrictions.

- The four variables $\{X, p_1, p_2, p_3\}$ in the primal LP are directly associated with the four restrictions in the dual problem, in a one-to-one relation; and vice-versa, the five primal restrictions are directly associated with the five dual variables $\{Y, q_1, q_2, q_3, q_4\}$.

- The coefficients for the primal variables in the objective function (1, 0, 0, 0) are moving to the right-hand side of the associated restrictions; and vice-versa, the primal right-hand side parameters (1, 0, 0, 0, 0) are moving to the objective function coefficients. In the matrix form, this can be easily done by swapping and transposing the associated vectors for the objective function's coefficients and right-hand side.

- The restrictions matrix for the primal problem originates the restrictions matrix for the dual problem through transposition.

The matrix form for the dual problem obtained in this manner follows:

$$\min \ [W] = \begin{bmatrix} 1 & 0 & 0 & 0 & 0 \end{bmatrix} \cdot \begin{bmatrix} Y \\ q_1 \\ q_2 \\ q_3 \\ q_4 \end{bmatrix}$$

subject to

$$\begin{bmatrix} 0 & 1 & 1 & 1 & 1 \\ 1 & -15 & -5 & -25 & -10 \\ 1 & -25 & -15 & -10 & -50 \\ 1 & -20 & 0 & -5 & -20 \end{bmatrix} \cdot \begin{bmatrix} Y \\ q_1 \\ q_2 \\ q_3 \\ q_4 \end{bmatrix} \geq \begin{bmatrix} 1 \\ 0 \\ 0 \\ 0 \end{bmatrix}$$

$$\begin{bmatrix} Y & q_1 & q_2 & q_3 & q_4 \end{bmatrix}^T \geq 0$$

Then the minimization LP for *Player-Two* is obtained by applying the algebraic form of the dual LP problem:

$$\min W = Y + 0q_1 + 0q_2 + 0q3 + 0q_4$$

subject to

Q) $\quad q_1 + \quad q_2 + \quad q_3 + \quad q_4 = 1$

A1) $Y - 15q_1 - 5q_2 - 25q_3 - 10q_4 \geq 0$

A2) $Y - 25q_1 - 15q_2 - 10q_3 - 50q_4 \geq 0$

A3) $Y - 20q_1 \qquad - 5q_3 - 20q_4 \geq 0$

The complementary properties for the optimal solution of both the dual problem and the primal problem apply. These properties can be observed by comparing the optimal solution for the primal LP problem,

$$\begin{bmatrix} X \\ p_1 \\ p_2 \\ p_3 \\ s_P \\ s_{B1} \\ s_{B2} \\ s_{B3} \\ s_{B4} \end{bmatrix}^* = \begin{bmatrix} 13 \\ 0.2 \\ 0.8 \\ 0 \\ 0 \\ 10 \\ 0 \\ 0 \\ 29 \end{bmatrix}, \text{ and the complimentary dual results } \begin{bmatrix} r_X \\ r_{A1} \\ r_{A2} \\ r_{A3} \\ Y_P \\ q_1 \\ q_2 \\ q_3 \\ q_4 \end{bmatrix}^* = \begin{bmatrix} 0 \\ 0 \\ 0 \\ 11 \\ 13 \\ 0 \\ 0.6 \\ 0.4 \\ 0 \end{bmatrix}.$$

8.5 DOMINANT STRATEGIES

From the payments table for the Zero-Sum Game with stable equilibrium, it can be seen that option A3 (bolded) is always worse for *Player-One* than option A2, whatever the choice of *Player-Two* (Table 8.29).

TABLE 8.29   Payments Matrix for Zero-Sum Game with Stable Equilibrium: Dominated Strategy A3

Payments *Player-One*	B1	B2	B3	B4
A1	15	5	25	10
A2	25	15	20	50
A3	20	0	5	20

As a result, option A2 dominates option A3, and *Player-One* will dismiss the third option and ponder only options A1 and A2.

With the same reasoning, it can be seen that option B2 is always better for *Player-Two* compared to the other options (B1, B3, and B4; bolded), independent of the choices *Player-One* makes (Table 8.30).

TABLE 8.30   Payments Matrix for Zero-Sum Game with Stable Equilibrium: Dominated Strategies B1, B3, and B4

Payments *Player-Two*	B1	B2	B3	B4
A1	−15	−5	−25	−10
A2	−25	−15	−20	−50
A3	*	*	*	*

Thus, option B2 dominates the other options for *Player-Two*, which causes him to dismiss them and ponder only option B2.

In this way, the set of results will be restricted, while stable equilibrium for the options pair (A2, B2) still is a suitable choice, as presented in Table 8.31.

TABLE 8.31 Payments Matrix for Zero-Sum
Game with Stable Equilibrium: Dominant Strategies

Payments Matrix	B1	B2	B3	B4
A1	*	(5,–5)	*	*
A2	*	**(15,–15)**	*	*
A3	*	*	*	*

Now, let us apply a similar approach for the dynamic interactions instance, remembering that it drives a cycle of losses and improvements, both for *Player-One* and *Player-Two*. From the payments table, once more it can be seen that option A2 dominates A3 (in bold), and whatever choice *Player-Two* makes, *Player-One* will dismiss A3 (Table 8.32).

TABLE 8.32 Payments Matrix for Zero-Sum
Game with Cycle: Dominated Strategy A3

Payments *Player-One*	B1	B2	B3	B4
A1	15	5	25	10
A2	25	15	10	50
A3	**20**	**0**	**5**	**20**

Similarly, it can be seen that option B2 dominates options B1 and B4 (in bold), and *Player-Two* will dismiss them independent of the choices made by *Player-One* (Table 8.33).

TABLE 8.33 Payments Matrix for Zero-Sum
Game with Cycle: Dominated Strategies B1 and B4

Payments *Player-Two*	B1	B2	B3	B4
A1	**–15**	–5	–25	**–10**
A2	**–25**	–15	–10	**–50**
A3	*	*	*	*

In this way, the game results will be restricted too, while the dynamic cycle of interactions for the options pair (A1–A2, B2–B3) still represent suitable choices as shown in Table 8.34.

TABLE 8.34 Payments Matrix for Zero-Sum Game with Cycle: Dominant Strategies

Payments Matrix	B1	B2	B3	B4
A1	*	(5,–5)	(25,–25)	*
A2	*	(15,–15)	(10,–10)	*
A3	*	*	*	*

8.5.1 Player-One: Dominant Strategies and LP

Integrating the mixed strategies approach with the LP formulation, and assuming that *Player-One* dismisses option A3, then the probability variable p_3 can be ignored. Also, assuming *Player-Two* dismisses options B1 and B4, then the associated restrictions can be ignored too. The updated LP formulation thus follows:

$$\max Z = X + 0p_1 + 0p_2$$

subject to

$$\text{P)} \qquad p_1 + \quad p_2 = 1$$
$$\text{B2)} \quad X \le \ 5p_1 + 15p_2$$
$$\text{B3)} \quad X \le 25p_1 + 10p_2$$

Note that

$$p_2 = 1 - p_1$$

and then the restrictions lines for the expected value can be obtained. From the B2-line, the restriction line will be:

$$X = 5p_1 + 15p_2$$
$$= 5p_1 + 15(1 - p_1)$$
$$= 5p_1 + 15 - 15p_1$$
$$= 15 - 10p_1$$

And from the B3-line, the restriction line will be:

$$X = 25p_1 + 10p_2$$
$$= 25p_1 + 10(1 - p_1)$$
$$= 25p_1 + 10 - 10p_1$$
$$= 10 + 15p_1$$

The intersection between the two restriction lines occurs when:

$$15 - 10p_1 = 10 + 15p_1$$
$$15 - 10 = 10p_1 + 15p_1$$
$$5 = 25p_1$$
$$p_1 = 5/25$$
$$p_1 = 0.2$$

And finally,

$$p_2 = 1 - 0.2 = 0.8$$

The optimal value for the maximization of expected gains X for *Player-One* is, from the B2-line,

$$X = 15 - 10 \times (0.2) = 15 - 2 = 13$$

or from the B3-line,

$$X = 10 + 15 \times (0.2) = 10 + 3 = 13$$

In addition, these values can be compared with the numeric results for the related LP (in Section 8.4), and they can also be observed in Figure 8.1.

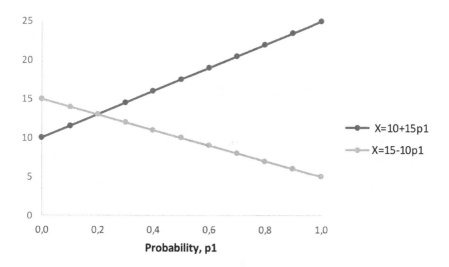

FIGURE 8.1   Expected gain X for *Player-One*: Variation with probability p_1.

8.5.2 Player-Two: Dominant Strategies and LP

Integrating the mixed strategies approach with the LP formulation, and assuming that *Player-Two* dismisses options B1 and B4, then the probability variables q_1 and q_4 can be ignored. Also, assuming *Player-One* dismisses options A3, then the associated restriction can be ignored too. The updated LP formulation thus follows:

$$\min W = Y + 0q_2 + 0q_3$$
$$\text{subject to}$$
$$\text{Q)} \qquad q_2 + \quad q_3 = 1$$
$$\text{A1)} \ \ Y \geq \ 5q_2 + 25q_3$$
$$\text{A2)} \ \ Y \geq 15q_2 + 10q_3$$

Note that

$$q_3 = 1 - q_2$$

and then the restrictions lines for the expected value can be obtained. From the A1-line, the restriction line will be:

$$Y = 5q_2 + 25q_3$$

$$= 5q_2 + 25(1 - q_2)$$

$$= 5q_2 + 25 - 25q_2$$

$$= 25 - 20q_2$$

And from the A2-line, the restriction line will come:

$$Y = 15q_2 + 10(1 - q_2)$$

$$= 15q_2 + 10(1 - q_2)$$

$$= 15q_2 + 10 - 10q_2$$

$$= 10 + 5q_2$$

The intersection between the two restriction lines occurs when:

$$25 - 20q_2 = 10 + 5q_2$$

$$25 - 10 = 20q_2 + 5q_2$$

$$15 = 25q_2$$

$$q_2 = 15/25$$

$$q_2 = 0.6$$

And finally,

$$q_3 = 1 - 0.6 = 0.4$$

The optimal value for the minimization of expected losses Y for *Player-Two* is, from the A1-line,

$$Y = 25 - 20 \times (0.6) = 25 - 12 = 13$$

or from the A2-line,

$$Y = 10 + 5 \times (0.6) = 10 + 3 = 13$$

These values can be compared with the numeric results for the related LP (in Section 8.4), and can also be observed in Figure 8.2.

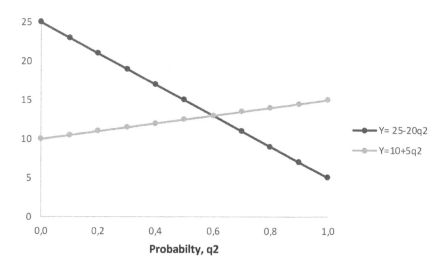

FIGURE 8.2   Expected loss Y for *Player-Two*: Variation with probability q_2.

8.6 CONCLUDING REMARKS

A brief discussion on game theory has been provided in this chapter. The key ideas associated with the two-player constant-sum game, zero-sum game, as well as mixed strategies and dominant strategies, have been introduced.

Game theory broadens the decision framework by addressing two decision-makers; in the LP framework, typically only one decision-maker is considered, but LP duality allows for the proper integration of a second decision-maker. As a result, LP complementary versions, both Max-Min and Min-Max, are presented for the games and instances under analysis.

A suitable generalization within the decision-making landscape considers multiple decision-makers, or even decision by group. The equilibrium stable point, or the dynamic cycle of alterations, can be also predicted through Differential Calculus; the approach is different, but can be aligned with LP duality.

However, considering the limitations of LP in light of data variability and alterations over time, as well as the different perspectives of each of the many decision-makers (and their attributes, beliefs, and values), the next chapter is dedicated to decision-making under uncertainty and expands decision-maker behavior beyond the pessimistic profile.

CHAPTER 9

Decision-Making Under Uncertainty

T HE MAIN TOPICS ON decision-making under uncertainty are presented, including important theoretical and practical issues. Advanced methods both for assisting the decision-making process and for modeling decision methods in uncertain conditions are addressed, and a comparison analysis is also developed. The comparison of alternatives using multiple criteria is crucial for the proper treatment of large and complex projects, as for example, the capacity expansion for a furniture factory. Decision-makers' real world decisions are molded through the ponderation of their attitudes, behaviors, beliefs, and values, and the appreciation of such decision-making processes requires advanced methods.

9.1 INTRODUCTION

Optimization methods usually assume that problem data, coefficients, and parameters are known without any doubts, uncertainty, or errors. But in reality:

- A system's behavior is not fully known, the related models are not complete, and they can change with time.

- The external variables and non-controllable factors can influence the outcome of each decision, sometimes in a crucial manner.

DOI: 10.1201/9781315200323-9

- The knowledge related to the variables that significantly act on a system (controllable or not) can assume a non-deterministic nature.

Decision-making under uncertainty occurs when it is not possible to associate any frequency or occurrence probability with the non-controllable variables. In this way:

- Experts can be used to establish alternative scenarios, which correspond to possible states of nature and related developments.

- In view of each scenario, each state of nature, it will be possible to select the most appropriate decision.

- The results space for the decision process can be represented in two dimensions (decision *vs.* scenario), usually in a matrix form.

An appreciation of the multiple attributes that affect a decision-maker's reasoning, either their behavior, beliefs, or values, is important for an accurate comparison of alternatives within uncertain contexts. In the next section, important criteria and decision-maker values are addressed; in Section 9.3, those criteria are illustrated by applying them to the capacity expansion of a furniture factory; in Section 9.4, a comparative analysis of the related outcomes is developed; and finally, in Section 9.5, the concluding remarks are discussed.

9.2 MULTIPLE CRITERIA AND DECISION-MAKER VALUES

In the real world, decision-making generally calls on multiple attributes, the decision-maker's values are questioned in face of delicate situations, and the complexity of problems usually requires enhanced methods to cope with uncertainty. The following approaches are commonly developed, in concordance with decision-maker behavior.

A. *Optimistic*

The decision-maker supposes the realization of the best possible scenario, in order to favor his situation; for example, it is assumed that the greatest gain, or the lowest cost, will occur.

This criterion supposes the decision-maker's optimistic behavior; he will select the alternative that assures the best result from the best possible scenarios. Addressing calculations, a maximization approach will be used to achieve the best gain (*e.g.*, Max-Max), while a minimization approach will be used to achieve the lowest loss.

B. *Pessimistic*

In the opposite sense to the optimistic approach, the decision-maker supposes the realization of the worst possible scenario; now it is assumed that the most unfavorable situation will occur, for example, the situation related to the highest cost, or the lowest gain.

In line with the criterion name, a pessimistic behavior for the decision-maker is assumed; he will select the alternative that assures the best result when considering the worst possible scenarios. Addressing calculations, a combination of maximization and minimization approaches (*e.g.*, Max-Min) will be used, in concordance both with the problem sense and the related data.

C. *Realistic*

This criterion balances between the two first criteria, *optimistic* and *pessimistic*, by introducing the optimism coefficient (α). The decision-maker defines his optimism level as a fractional number between 0 and 1, where: $\alpha = 1$ corresponds to a fully optimistic decision-maker, and $\alpha = 0$ corresponds to a fully pessimistic one.

From this point of view, the third criterion generalizes the two first criteria, referred to in prior points A and B; or from another point of view, the *realistic* criterion can be seen to generate the *optimistic* and *pessimistic* criteria, when α takes the value 1 or 0, respectively.

This criterion represents a linear combination of the optimistic and pessimistic criteria and is based on a more realistic reasoning. The decision-maker will select the alternative that assures the best result, while the probabilities associated with the possible scenarios are weighted through the optimism coefficient, α.

Therefore, it is also of interest to develop a sensitivity analysis on this weighting coefficient. Calculations will be used that follow a combination of maximization or minimization searches (*e.g.*, or Max; or Min), in concordance both with the problem sense and the related data (respectively, or gains; or costs).

D. *Uniform Probability*

This criterion supposes a uniform distribution of probabilities, that is, equal probabilities are assumed for each possible scenario. The decision-maker selects the alternative with the best result for the equal-distributed and uniform appreciation of scenarios.

The *uniform* criterion, also known as the Laplace criterion, appreciates the usual arithmetic mean or average, and a combination of maxima or minima searches (*e.g.*, or Max; or Min) will be used in concordance both with the problem sense and data, as discussed previously.

E. *Loss of Opportunity*

It is also a common procedure for the decision-maker to build a matrix with the opportunity costs; that is, since he does not know what the future holds, then the best alternative for each possible scenario is retained and the loss when not selecting such alternative is calculated.

This criterion is based on minimizing the opportunity costs. With this approach, opportunity cost is the loss or the benefit that could have been enjoyed if the best alternative was chosen; it is also known as "loss of opportunity" or "cost of opportunity". For each possible scenario, the decision-maker evaluates the difference between the results corresponding to the best alternative and each one of the other alternatives under consideration, and a matrix with the opportunity losses is obtained; then he will select the alternative leading to the minimum opportunity cost for all the possible scenarios.

The calculations follow through a combination of Min-Max approaches, while the value of perfect information is also appreciated; in this way, higher values for the opportunity costs can lead to additional studies that can reduce the existing uncertainty (*e.g.*, market studies, customer panels).

9.3 CAPACITY EXPANSION FOR THE FURNITURE FACTORY

Capacity expansion is significant for many industries, since it is characterized by the allocation of substantial resources. An investment in capacity expansion is usually planned with long payback periods and provides economies of scale in line with the technology enhancements (for example, in heavy industry, metallurgy, petrochemical, pharmaceutical).

Typically, capacity expansion is a problem under uncertainty that includes, among others, these general attributes:

- Obsolescence (*e.g.*, in the electronics industry) and deterioration (*e.g.*, in heavy machinery), leading to increased maintenance and operating costs.

- Multiple types of technology (*e.g.*, renewable energies) and discretization of equipment capacities (*e.g.*, volume, tonnage, quantities, or rates).

- Economies of scale, integrating both fixed-charge and variable costs, and finance topics.

The strategy for capacity expansion does not have to follow the attributes associated with the minimum investment. In fact, although the initial objective may be to satisfy the increasing demand for products with the smallest investment possible, another option is to define the best use for the existing facilities, within the pre-fixed capacities.

In addition, uncertainty in product demand may cause two types of deviations: i) excess of capacity, which represents capital resources that are not efficiently used and that may affect future expansions, and ii) not-satisfied demand, with consequences at the service level, and for shortages or penuries.

9.3.1 The Furniture Factory's Capacity Expansion Problem

The Furniture Factory managers aim to introduce a new design line for furniture production, and the following alternatives are identified:

A1: Construct a new factory, with up-to-date technology.

A2: Expand the existing factory, with the same type of technology.

A3: Subcontract production overloads to other factories, due to the demand for the new design furniture.

A4: New production shifts (*e.g.*, nights, weekends) to cope with the new demand, while still using the same factory.

In terms of the demand for the new design furniture, the most significant scenarios for the furniture market are:

S1: Very good demand, due to the success of the new design furniture.

S2: Good demand, with good acceptance of the new design.

S3: Satisfactory demand for the new design by customers.

S4: Fair demand, with customers not connected to the new design.

S5: Poor demand, with customers refusing the new design furniture.

The gains matrix (in thousands of euros) is presented in Table 9.1.

Apply the criteria described in Section 9.2 to evaluate the five scenarios (S1–S5), and select one of the four alternatives (A1–A4).

TABLE 9.1　Gains matrix for the Furniture Factory's Capacity Expansion

Gains (× 1000 €)	S1 Very Good Demand	S2 Good Demand	S3 Satisfactory Demand	S4 Fair Demand	S5 Poor demand
A1: Construct	355	180	60	−150	−300
A2: Expand	315	190	80	−100	−200
A3: Subcontract	170	90	40	−25	−50
A4: New Shifts	85	45	20	−5	−10

a) The *Optimistic* Behavior

The realization of the best possible scenario is assumed, namely, the state of nature associated with the greatest gain will occur.

Supposing the decision-maker's optimistic behavior, a maximum search will be used to select the best gain. The first maximization step within this Max-Max criterion, the section of the best gain, is bolded at each line in Table 9.2.

TABLE 9.2　Gains Matrix for the Max-Max Criterion: First Step

Optimistic	Very Good Demand	Good Demand	Satisfactory Demand	Fair Demand	Poor Demand
Construct	**355**	180	60	−150	−300
Expand	**315**	190	80	−100	−200
Subcontract	**170**	90	40	−25	−50
New Shifts	**85**	45	20	−5	−10

For all the alternatives under consideration, A1–A4, the best state of nature is related to the expectation of a very good demand for the new furniture. Hereafter, the second maximum search is selecting alternative A1, indicating that the maximum gain is 355, and it is associated with the construction of a new factory, as presented in Table 9.3.

TABLE 9.3   The Second Step for
the Max–Max Criterion

Optimistic	Maximum Gain
Construct	**355**
Expand	315
Subcontract	170
New Shifts	85

b) The *Pessimistic* Behavior

The realization of the worst possible scenario is assumed, namely, the state of nature associated with the lowest gain will occur.

Supposing the decision-maker's pessimistic behavior, first, a minimum search will be used to select the lowest gain. The first minimization step within this Max-Min criterion, the lowest gain, is bolded at each line in Table 9.4.

TABLE 9.4   Gains Matrix for the Max-Min Criterion: First Step

Pessimistic	Very Good Demand	Good Demand	Satisfactory Demand	Fair Demand	Poor Demand
Construct	355	130	60	−150	**−300**
Expand	315	190	30	−100	**−200**
Subcontract	170	90	40	−25	**−50**
New Shifts	35	45	20	−5	**−10**

Now, for all the alternatives under analysis, the worst state of nature is related to a poor demand, with customers refusing the new furniture. Hereafter, the decision-maker's second step is selecting alternative A4, indicating the maximum value from the minima; or better, the minimum value from the greater losses, −10, associated with the introduction of new production shifts, as presented in Table 9.5.

TABLE 9.5 The Second Step of
Max-Min Criterion

Pessimistic	Minimum Gain (Greater Loss)
Construct	−300
Expand	−200
Subcontract	−50
New Shifts	**−10**

c) A *Realistic* Behavior

As introduced in Section 9.2, the decision-maker assumes a more realistic behavior, weighting between *optimistic* and *pessimistic* behavior by introducing the optimism coefficient, α. The decision-maker defines the optimism level as a fractional number; in this case, let it be $\alpha = 0.5$. The decision-maker is assumed to be half *optimistic* and half *pessimistic*. The weighted gains for each alternative are presented in Table 9.6.

TABLE 9.6 Gains Matrix for the *Realistic* Criterion: First Step

Realistic	Very Good Demand	Good Demand	Satisfactory Demand	Fair Demand	Poor Demand	Ponderation (Weighting, α)
Construct	355	130	60	−150	−300	27.5
Expand	315	190	30	−100	−200	57.5
Subcontract	170	90	40	−25	−50	60,0
New Shifts	85	45	20	−5	−10	37.5

For example, the weighted gain $W(A2)$ of alternative A2-*Expand* follows:

$$W(A2) = (0.5) \times 315 + (1 - 0.5) \times (-200)$$

$$= 157.5 - 100$$

$$= 57.5$$

And the weighted gain $W(A3)$, corresponding to alternative A3-*Subcontract* is:

$$W(A3) = (0.5) \times 170 + (1 - 0.5) \times (-50)$$

$$= 85 - 25$$

$$= 60$$

The decision-maker will select the alternative that assures the best weighted gain. Therefore, a maximum search is selecting alternative A3, indicating that the maximum weighted gain is 60, and it is associated with subcontracting the production overloads to other factories, as presented in Table 9.7.

TABLE 9.7   The Second Step for the *Realistic* Criterion

Realistic	Ponderation (Weighting, α)
Construct	27,5
Expand	57,5
Subcontract	60,0
New Shifts	37,5

The difference between the two best alternatives for the *realistic* criterion is very small. As a result, a sensitivity analysis on the optimism coefficient, α, is pertinent, and such analysis will be developed in the next section.

d) The *Uniform* Approach

The *uniform* criterion supposes that equal probabilities are assumed for each possible scenario, that is, each one of the five states of nature present an equal probability of occurrence: $1/5 = 0.20$. The expected gains for each alternative are presented in Table 9.8.

TABLE 9.8   Gains Matrix for the *Uniform* Criterion: First Step

Uniform Probability	Very Good Demand	Good Demand	Satisfactory Demand	Fair Demand	Poor Demand	Expectation
Construct	355	180	60	−150	−300	29
Expand	315	190	30	−100	−200	57
Subcontract	170	90	40	−25	−50	45
New Shifts	35	45	20	−5	−10	27

For example, the expected gain $E(A2)$ of alternative A2-*Expand* follows:

$$E(A2) = 0.2 \times 315 + 0.2 \times 190 + 0.2 \times 80 + 0.2 \times (-100) + 0.2 \times (-200)$$

$$= 63 + 38 + 16 - 20 - 40$$

$$= 57$$

And the expected gain $E(A3)$, corresponding to alternative A3-*Subcontract* is:

$$E(A3) = 0.2 \times 170 + 0.2 \times 90 + 0.2 \times 40 + 0.2 \times (-25) + 0.2 \times (-50)$$

$$= 34 + 18 + 8 - 5 - 10$$

$$= 45$$

The decision-maker will select the alternative that assures the best expected gain. Then, a maximum search is selecting alternative A2, indicating that the maximum expected gain of 57 is associated with expanding the furniture factory, as presented in Table 9.9.

TABLE 9.9   The Second step for the *Uniform* Criterion

Uniform Probability	Expectation
Construct	29
Expand	57
Subcontract	45
New Shifts	27

e) The *Loss of Opportunity* Approach

The decision-maker prepares a matrix with the opportunity costs; for that, the best alternative for each possible scenario is selected through a maximum search per column. For the selected alternatives, bold lettering is utilized in Table 9.10.

TABLE 9.10   Gains Matrix for the *Opportunity* Criterion: The Preparation Step

Gains	Very Good Demand	Good Demand	Satisfactory Demand	Fair Demand	Poor Demand
Construct	**355**	180	60	-150	-300
Expand	315	**190**	**80**	-100	-200
Subcontract	170	90	40	-25	-50
New Shifts	85	45	20	**-5**	**-10**

The best alternative for each state of nature is A1-*Construct*, with a gain of 355 for very good demand; A2-*Expand*, with gains of 190 and 80, for good demand and satisfactory demand, respectively; and A4-*New Shifts*, with of gains –5 and –10 corresponding, respectively, to fair demand and poor demand.

Then, at each column, the decision-maker evaluates the difference between the best alternative and all of the other alternatives. A matrix with the opportunity costs is obtained, as presented in Table 9.11.

TABLE 9.11　Costs Matrix for the *Opportunity* Criterion: First Step

Opportunity Costs	Very Good Demand	Good Demand	Satisfactory Demand	Fair Demand	Poor Demand	Maximum of Opportunity Losses
Construct	0	10	20	145	290	290
Expand	40	0	0	95	190	190
Subcontract	185	100	40	20	40	185
New Shifts	270	145	60	0	0	270

For example, alternative A3-*Subcontract* presents the following gains for the five states of nature, in descending order from very good demand to poor demand: 170, 90, 40, –25, and –50. The opportunity costs corresponding to the alternative A3-*Subcontract* are calculated through the following differences:

o For very good demand, the gain of alternative A3 is 170; the difference to alternative A1-*Construct* with gain 355 thus is 185.

o For good demand, A3 gain is 90; the difference to alternative A2-*Expand*, with gain 190 thus is 100.

o For satisfactory demand, A3 gain is 40; the difference to alternative A2-*Expand*, with gain 80 is 40.

o A3 gain is –25 in case of fair demand; the difference to alternative A4-*New Shifts* with gain –5 thus is 20.

o A3 gain is –50 in case of poor demand; the difference to alternative A4-*New Shifts*, with gains –10 thus is 40.

o Therefore, the maximum loss of opportunity for alternative A3-*Subcontract* is 185, as indicated in the far-right column of Table 9.11.

The decision-maker will select the alternative leading to the minimum opportunity cost for all the possible scenarios. Then, a minimum search is selecting alternative A3, indicating that the minimum loss of opportunity is 185, and it is associated with subcontracting the production overloads to other factories, as presented in Table 9.12.

TABLE 9.12   The Second Step for the *Opportunity* Criterion

Opportunity Costs	Min-Max of Opportunity Losses
Construct	290
Expand	190
Subcontract	**185**
New Shifts	270

These values for the opportunity costs can indicate upper bounds for the expected value of perfect information; namely, market studies or customer panels that can contribute to reduce uncertainty and that are sufficiently lower than the opportunity costs.

9.4 A COMPARISON ANALYSIS

The recommendations and gains related to all the five methods are indicated in Table 9.13, allowing this way a better comparison between them.

TABLE 9.13   Recommendations and Gains for the Different Methods

Method	Recommendation	Gains
Optimistic (Max-Max)	Construct	355
Pessimistic (Max-Min)	New Shifts	−10
Realistic (Weighting)	Subcontract	60
Uniform Probability (Expectation)	Expand	57
Loss of Opportunity (Opportunity Cost)	Subcontract	185

It can be observed that:

- The *optimistic* criterion using the Max-Max approach presents the best gain and the bolder recommendation (A1-*Construct*), while the *pessimistic* criterion using the Max-Min approach presents the best of the worst gains and the most conservative recommendation (A4-*New Shifts*).

- In-between, the *realistic* criterion uses the optimism coefficient, α, and selects the best of the weighted gains and an intermediate recommendation (A3-*Subcontract*).

- In addition, while the *uniform* criterion uses the maximum of expected gains to select the other intermediate recommendation (A2-*Expand*), the *opportunity cost* approach is selecting the same recommendation (A3-*Subcontract*) as the *realistic* criterion.

As mentioned previously, it is of interest to develop a sensitivity analysis on the optimism coefficient, α.

- The *realistic* criterion generalizes the *optimistic* and *pessimistic* criteria, in this sense it coincides with these two methods when α takes the value 1 or 0. In fact, the *realistic* approach reduces to the best gain and the bolder recommendation (A1-*Construct*) in the first case, and to the most conservative recommendation (A4-*New Shifts*) in the latter, as presented in Table 9.14.

TABLE 9.14   Sensitivity Analysis on the Optimism Coefficient, α.

Optimism Coefficient (α)	Gain	Recommendation
0	−10,0	New Shifts
0,1	−1,0	New Shifts
0,2	9,0	New Shifts
0,3	19,0	New Shifts
0,4	38,0	Subcontract
0,5	60,0	Subcontract
0,6	109,0	Expand
0,7	161,0	Expand
0,8	224,0	Construct
0,9	290,0	Construct
1	355,0	Construct

- In fact, the *realistic* approach can be seen as a linear combination of the *optimistic* and *pessimistic* approaches, and the intermediate gains and related recommendations can be found too. The intermediate alternative A2-*Expand* is selected for the range 0.6–0.7, while the alternative A3-*Subcontract* is selected for the range 0.4–0.5.

- The solid line for the median value, $\alpha = 0.5$, which drives an intermediate gain associated with a mid-term recommendation (A3-*Subcontract*), corresponds to the instance treated in the previous section.

A graphic representation is a very common tool to improve the insight about the optimism coefficient's variation, from 0 to 1 (Figure 9.1).

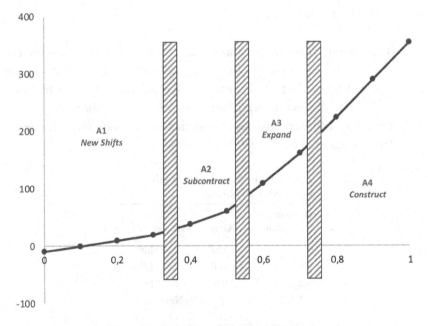

FIGURE 9.1　Sensitivity analysis on the optimism coefficient, α.

The criteria, as well as the related methods, present both strengths and weaknesses. Applying one of these methods depends on the organization's policy in face of uncertainty, which is closely connected with the decision-maker's behavior, beliefs, and values. Table 9.15 lists the strong and weak points for each of the methods described.

TABLE 9.15   Decision-Making in Uncertain Conditions.

Method	Strengths	Weaknesses
Optimistic (**Max-Max**)	Positive behavior.	High level of risk assumed.
Pessimistic (**Max-Min**)	Lowest level of risk assumed.	Excludes the possibility of very high gains.
Realistic (**Ponderation**)	More realistic.	A method for specifying the optimism coefficient is necessary.
Uniform Probability (**Expectation**)	Good if very few information is available.	Equal-distribution assumption needs additional support, theoretically or empirically.
Loss of Opportunity (**Opportunity Cost**)	A minimum for the opportunity costs is defined.	Excludes the possibility of very high gains.

9.5 CONCLUDING REMARKS

In this chapter, important topics for decision-making under uncertainty are discussed, including both basic notions of interest and the diverse criteria that best fit the decision-maker's profile. The appreciation of the decision-maker's multiple attributes, beliefs, and values is key, since these multiple factors encapsulate the decision process and mold the final decision.

The decision-maker can select different viewpoints, such as optimistic, pessimistic, or a more realistic approach; in addition, probability distributions (*e.g.*, uniform distribution) can be selected to allocate frequency probabilities to each nature state, or even utility or opportunity cost can be obtained through an economic approach. In this way, the strengths and weaknesses of each method can also be aligned with the organization's values, as well as the implementation strategy for the organization's mission. The appreciation of multiple attributes, as well as the values structure, values subordination, and values hierarchy, inspire a diversity of methods for Multi Criteria Decision Aiding/Making (MCDA/M).

The multiple attributes comparison of alternatives thus is crucial, and an illustrative example featuring the capacity expansion for the Furniture Factory was presented. In fact, capacity expansion planning is significant for many industries and is characterized, in general by the allocation of substantial resources; investments with long payback periods; searching for economies of scale; or selecting the technology types to implement.

Therefore, demand uncertainty in capacity expansion may cause two types of deviations: i) capacity excess, which represents capital resources that are not efficiently used, and that may also affect future expansions, and ii) non-satisfied demand, with the usual consequences associated with low service levels, by imposing shortages or penuries. These two types of deviations can be calculated and utilized to penalize the objective value, in order to promote robustness both for solutions and models, as presented in the next chapter dedicated to Robust Optimization.

CHAPTER 10

Robust Optimization

T HIS CHAPTER CONTINUES THE treatment of uncertainty and focuses
on robust optimization (RO); the decision-maker is assuming and
developing different scenarios in line with uncertain environments,
which requires stochastic treatments beyond the arithmetic mean or
expected value. Stochastic programming (SP) and its related attributes
for decision-making are presented, while the generalization of deterministic models to the stochastic framework that drive qualitative or
quantitative improvements are noted. In addition, robustness, both on
models and solutions, is achieved through RO approaches; namely, RO is
able to promote both the assessment and treatment of risk. A case study
is described, including pertinent RO procedures associated with typical
economic estimators and industry-based parameters, and tables summarizing the approaches and methods described throughout the book
are also presented.

10.1 INTRODUCTION

Uncertain or incomplete data is common in capacity expansion problems and other complex optimization problems that consider long-range
periods. The optimal LP solutions are usually treated in a reactive way
through sensitivity analysis, however it can be done proactively through
SP formulations. The stochastic linear programming (SLP) framework
promotes robustness, both on solutions or models.

DOI: 10.1201/9781315200323-10

The usual sensitivity analysis results in a reactive study on LP optimality, as it only looks at the impact of data uncertainty on the optimal values for decision variables. Diverse and complementary analysis are often assumed, either by the mean average value (*e.g.*, the expected value problem, EVP) or by the worst-case scenario (*e.g.*, the *pessimistic* decision-maker). Note the first approach can select solutions with high deviation, while the latter can induce conservative choices and more expensive solutions. These procedures correspond to a reactive approach on LP optimality that does not allow control measures, so it is necessary to adapt and enhance the decision procedures.

On the other hand, the SP approach on optimality is proactive. The recommendations obtained from the SP model can be adjusted to the data instances in the second phase, and thus decision-makers can rely on the flexibility of the resources or control variables. However, only the expected value for the SLP objective function is optimized, the variability of the solution is ignored. In addition, SP also disregards the decision-maker's preferences when faced with risk.

Indeed, the robust optimization (RO) framework promotes the assessment and risk treatment in conformity with the decision-maker's profile, namely, specific economic estimators and technical parameters can be introduced to better address uncertainty. With RO, improvements in the robustness of models and solutions are achieved.

To better address this topic, a case study addressing capacity expansion is briefly presented. The case study addresses a typical industry-based problem, the *design and scheduling batch processes* (DSBP) problem, while the meaning of more complex mathematical formulations is described. Within the general problem for capacity expansion, and in particular when addressing the capabilities of a chemical process network, some important operational aspects follow:

- Accurate forecasts for both product demands and prices, as well as for raw and intermediate materials, allow to better outline the states of nature.

- The capacity of each batch process, in each period of the time horizon, belongs to the design variables set, since the equipment sizes available for selection are not associated with the uncertain demand for each product.

- However, for each time period, variables associated with product sales, flows in the process network, and material purchase, are all integrated into the control variables set; thus, in the second phase, uncertainty is introduced through the forecasted scenarios set, and the associated probabilities too.

The capacity expansion plans are carried out with a distant insight to the future, but in the project phase only the first step is compromised. Any deviations to the initial decision will be subject to further corrections, due to the successive re-appreciation of the initial assumptions. It is possible to estimate scenario probabilities and calculate penalty amounts in case of any deviations; for this, parametric analysis of sensitive decisions can take place, and the opportunity costs method can be useful.

In general, the process network system shall respond to unpredictable and persistent deviations from average operating conditions, otherwise the system would be inefficient or even inoperable.

- Each network subsystem or unit process shall be prepared to respond, typically, both to average service needs and to reasonable fluctuations, or deviations, around those average conditions.

- To provide some comfort to decision-makers, over-sizing appeared empirically; with the developments of statistical decision theories, such empiricism is being mitigated and partially removed.

With a similar point of view, diverse theoretical developments addressing optimization tools and decision-making approaches are presented in this book. With these tools in mind, decision-makers can certainly face their delicate, large, and complex challenges with more confidence. A summary of the introductory tooling, either approaches or methods focused on in the initial part of this book, is presented in Table 10.1.

TABLE 10.1 Decision Support with Introductory Tooling

Approaches/ Methods	Optimization Tools	Decision-Making
Introductory Approaches	Introductory tools: trial-and-error; additivity; proportionality; and computational support.	First notes on decision processes. One single decision-maker is assumed, as well as one alternative selected using one and only one attribute.
Linear Algebra (LA)	Simple solution methods can efficiently solve problems based on linear systems of equations (LSE).	LA presents important shortcomings (*e.g.*, non-integer or non-positive solutions; slacks not assumed in LSE; infinite solutions and infeasible LSE).
Linear Programming (LP Basics)	LP broadens results space, including slack variables, dual values, and the treatment of multiple RHS.	LP improvements are both qualitative and quantitative, adding degrees of freedom, and the treatment of sensitive factors.
LP Duality	Duality complements and enlarges LP to the non-feasible region. Duality concepts are of great importance, *e.g.*, primal-dual complementary relations, or dual economic insights.	A complementary, second decision-maker is assumed. The alternatives space is enlarged with the resources utilization's insight. LP duality is important for advanced approaches (LP optimality analysis, calculus optimization, game theory).
Calculus Optimization	A generalization approach is initiated with a simple optimization problem. Lagrange multiplier as partial derivative, and related relaxation methods.	Optimization calculus enhances knowledge for decision-makers, either in STEM fields or economics and enterprise sciences (game theory, various types of economic analysis). Liaison between LP dual values and Lagrange multipliers.

The generalization of deterministic models toward the stochastic framework is driving both qualitative and quantitative improvements. A summary of the more advanced optimization approaches and methods treated in this book is presented in Table 10.2. In general, the following are observed:

- Objective function and feasible space of solutions are successively enlarged, assuming quantitative improvements.

TABLE 10.2   More Advanced Optimization Approaches for Decision-Making

Approaches/ Methods	Optimization Tools	Decision-Making
LP Optimality Analysis	Sensitivity analysis is systematically developed. Parametric analysis is mapping alterations to the optimal solution. Multiple optimal solutions are identified.	Decision-makers gain access to new information sources. New insights can enhance the response to external fluctuations or other uncertainty factors. The alternatives space is enlarged, adding multiple optimal solution combinations.
Integer Linear Programming (ILP)	ILP assumes variable integrity and properly includes binary variables. Qualitative improvements by constraining the results space, at the cost of quantitative bounding, since *branch-and-bound* follows a reduction approach.	Decision-makers obtain solutions that are directly translated to the real world. Managers gain better modeling capabilities (non-proportionality in LP; logic and semi-logic relations; contingency decisions; fixed charge costs). Decision-makers use ILP very often, even for large and complex problems.
Game Theory	Game theory broadens the decision framework by addressing two decision-makers. Key notions are related to the two-player Constant-Sum Game and Zero-Sum Game. Complementary LP formulations are developed, both Max-Min and Min-Max.	The existence of a second decision-maker is assumed. Autonomously, *Player-Two* is selecting pertinent alternatives and impacting system results, assuming a *pessimistic* behavior. The treatment of both mixed strategies and dominant strategies is an important enhancement too.
Decision Under Uncertainty	Limitations of optimization methods in face of data variability and alterations over time. Comparison of alternatives using multiple criteria is key for the proper treatment of large and complex projects.	Expanding the decision-maker's behavior beyond the *pessimistic* profile (*e.g.*, *optimistic, realistic, probabilistic*). Including perspectives of each one of the decision-makers (attributes, beliefs, and values). Multiple factors impact the decision process and the final decision.

(Continued)

TABLE 10.2 (CONTINUED)   More Advanced Optimization Approaches
for Decision-Making

Approaches/ Methods	Optimization Tools	Decision-Making
Robust Optimization (RO)	RO extends stochastic treatment beyond EVP, the expected value problem. Robustness both on models and solutions can be achieved. RO allows risk treatment, economic estimators and technical parameters can be adjusted.	Generalizing optimization models to the stochastic framework is driving either qualitative or quantitative improvements for decision-makers. Decision-makers develop multiple scenarios in line with the uncertain environment. Broadening the decision process in two phases, or multiple phases, allows proactive actions on data variability and fluctuations.

- Continuous evolution with successively extended models, also assuming qualitative improvements for decision approaches.

- Evolution of solution methods too, with approximation procedures being required (*e.g.*, decomposition methods, Lagrangean relaxation).

- Among other approaches within the multi-criteria decision aiding/ making (MCDA/M) area, the RO importance is widely recognized.

10.2 NOTES ON STOCHASTIC PROGRAMMING

Stochastic programming (SP) addresses uncertainty, which is usually represented through probabilistic scenarios. The related attributes for SP-based solutions are revisited in this section, including probabilistic estimators (*e.g.*, expectance, variance, standard deviation) and the recourse phase for two-stage decision-making (or even multiple stages).

When assuming an uncertain environment with multiple states of nature, different scenarios are developed to deal with such uncertainty, and the optimization problem thus requires stochastic treatment. Typically, the stochastic objective aims at optimizing the expected value for the deterministic objective, while the sum of all the scenario probabilities is 1.

For example, in the capacity expansion problem, the net gain is updated to the current moment, ξ (net present value, NPV), and each one of the random scenarios (index r) is weighted by the associated probability, *prob*:

$$\max \Phi = \sum_{r=1}^{NR} \text{prob}_r \xi_r = E(\xi) \tag{10.1}$$

Counte-intuitively, the SP solution for the capacity expansion problem can present a lower-value for the expected investment cost, while over-capacity can occur for the majority of demand scenarios. In fact, SP formulations can take advantage of economies of scale, but these expansion plans are very sensitive to demand uncertainty and over-evaluating low probability scenarios. Thus, high demand scenarios with low probabilities can drive over-sizing solutions; this occurs because penalizing slack capacities is not assumed, otherwise such excess capacities could be mitigated.

For that, the decision-maker in the capacity expansion problem can adopt the two-stage SP (2SSP) approach, which explicitly considers two decision phases:

- In the first stage, the capital and investment decisions must be taken, "here-and-now"; then, the project variables are obtained. This phase is also known as the project stage.

- In the second stage, the uncertainty is introduced through the scenarios set and associated probabilities; in this way, control variables are obtained with a probabilistic attribute. This phase is also named the recourse stage.

As mentioned, the 2SSP modeling framework assumes two phases, where two different types of decision variables are observed. Namely:

- Project or design variables: these variables are related to the structural and deterministic constraints of the problem, and their calculation is not dependent on uncertain parameters.

- Control or recourse variables: these variables can be adjusted accordingly to the uncertain parameters at each scenario, but the respective values will depend on design variables that are already defined in the first stage.

In the two-stage approach, decision-makers define the problem variables in two phases: first, the project variables whose values are decided *a priori*, and from which the uncertain parameters are revealed; then, second, the recourse or control variables are presented as random variables, due to the associated uncertainty. As discussed, the 2SSP results undergo a corrective phase, the second or recourse stage, which minimizes the total amount of the design costs in the first phase with the expected value of the resource costs resulting from the random variables in the second phase.

10.3 ROBUSTNESS PROMOTION ON MODELS AND SOLUTIONS

A robust decision occurs when the corresponding cost for the particular scenario remains close to the optimal expected value for all scenarios. Usually, the variance of the scenario-dependent costs is constrained through the restriction of the expected deviation, so robustness is reinforced by penalizing the sensitivity of the objective function value to uncertain parameters.

Solution robustness implies the optimal solution is also of good quality when facing the entire set of uncertain scenarios. From another point of view, model robustness specifically refers to the optimal solution, which should not present significant losses due to the stochastic intrinsic variation.

For example, for production planning with uncertain demand for a set of goods or products, a robust solution must present small values for the non-satisfied demand, as well as for excess goods. For the capacity expansion problem, RO allows obtaining expansion plans characterized by:

- Solution robustness: occurs when, for any instance of the demand scenarios, the solutions set that are obtained remain close to the optimal solution (and the expected value). For that, it is important to constrain solution variability (*e.g.*, through variance, or absolute deviation) by penalizing it in the second phase's objective function; in this way, the optimal solution in RO presents good quality for all of the scenarios or states of nature under consideration.

- Model robustness: considering all the scenarios set, the optimal solution presents a restriction in over-capacity and the unmet demand is diminished, since the associated estimators are penalized; consequently, in the capacity expansion problem with uncertainty in demand forecasts, the robust model should not present high values for unmet demand, nor for capacity excess.

The parameters for the stochastic optimization problem, with robustness promotion, are:

Φ—stochastic objective function.

δ—set of random scenarios, index r.

λ, w—penalty parameters.

ξ—net present value (NPV).

$E(\xi)$, dev(ξ)—expected value and standard deviation for NPV.

x—design variables.

y_r—control variables, for each scenario r.

z_r—error or deviation from infeasibility, for each scenario r.

The objective function for the RO problem aims at maximizing the NPV expected value, subtracted by a penalty term associated with deviations from the expected value at each scenario, as well as subtracting a penalty for deviation in the modeling (*e.g.*, over-production, or under-utilization of capacities):

$$\max \ \Phi = E(\xi) - \lambda . \ E\left[\operatorname{dev}^2(\xi)\right] - w \cdot E(z_r^2) \tag{10.2}$$

This robust relation for the expected NPV's maximization is subject to restrictions that are independent from the multiple states of nature, that is, restrictions that are not affected by the probabilistic scenarios:

$$A \ x = b \tag{10.3}$$

In addition, restrictions depending on probabilistic scenarios are integrated too. The satisfaction of these scenario-dependent restrictions can be relaxed, being the extension of the corresponding deviation or infeasibility, z_r, measured in the second phase:

$$B_r \ x + C_r y_r + z_r = 0, \quad \forall r \tag{10.4}$$

Then, for penalization purposes, the infeasibility at each random scenario, r, is weighted in the third term of the stochastic objective function at the relation shown in Equation 10.2. In this manner, RO solutions can be directly compared to SP, being those SP solutions trivially calculated by taking zero for all the penalty parameters in the RO. That is, when the penalty parameters in the relation shown in Equation 10.2 take zero, $\lambda = w = 0$, then the relation in Equation 10.1 for SP is obtained.

Through the objective function, the probability distribution that describes the data uncertainty is distorted, which translates into the spread of uncertainty throughout the long-range period, with expansion and contraction of the range of probabilities, according to the sensitivity of each parameter under analysis. In any case, finance effects of uncertainty are mitigated by the decreasing exponential function in the NPV calculations, so the distant future loses comparative weight in today's decision-making.

When comparing SP and RO, two main topics are relevant:

- The first distinction is presented in the objective function, where RO controls the solution variability, while only the expected value is optimized in SP.

- Another differentiating issue is the treatment of restrictions; namely, an SP instance is not feasible when a combination of design variables (first phase) cannot be found that allow control variables (second phase) to satisfy the restrictions for all the scenarios (the complete recourse); however, such contingency is explicitly addressed in RO, by introducing relations (deviations and associated penalties) that allow solutions that violate restrictions with low severity.

However, some limitations occur within the RO scope regarding the specification of penalty parameters, and no means are provided to select scenarios and associated probabilities. The latter issue also occurs in SP, although these scenarios would constitute only a set of possible instances for the real problem.

In addition, the need to carry out a high number of iterations for the RO problem must also be taken into account in order to obtain sufficient information on the optimality for the robust solution before making the final decision, as it is usual in SP methods.

- The solution speed in RO can represent a barrier, the importance of which must be duly taken care of. Otherwise the solution cycles may drag unreasonably, for example, for integrity restrictions (*e.g.*, ILP).

- Indeed, the solution speed of sub-problems is a key attribute requiring attention to see if the SP is suitable for decomposition methods, either by scenarios or time periods. This situation often occurs in LP and ILP, and SLP extends the static LP onto the stochastic framework.

Note the SLP solution procedures can follow the common methods for deterministic LP. However, LP approximations do not insure the SLP's optimality in face of multiple states of nature, because uncertainty is expressed through different probabilistic scenarios.

10.4 MODELS GENERALIZATION ONTO ROBUST OPTIMIZATION

In this section, important aspects about the generalization of deterministic models onto the stochastic and robust frameworks are presented. For that, an RO case study is described, namely the *design and scheduling of batch processes* (DSBP), while operational aspects of interest are also introduced.

The deterministic DSBP problem simultaneously addresses the design and scheduling of a multi-product flowshop, while assuming multiple processes at each production stage (multiple machines), single product campaigns (SPC), and zero wait policy (ZW). The current approach allows the generalization of:

i) The deterministic and single-period model onto a stochastic framework with robustness.

ii) The single time period (static) onto a multi-period (dynamic) horizon, while optimizing the robust NPV associated with the investment in batch processes, which are formulated with discrete sizes (integer variables).

In this way, let the Robust-DSBP problem be the enlarged version of the DSBP problem, addressing multiple machines and SPC policy. Other attributes and data of specific interest for the Robust-DSBP problem follows:

Index and Sets

M	Number of stages j.
NC	Number of components or products i.
$NP(j)$	Number of processes $p(j)$ per stage.
NR	Number of discrete scenarios r.
$NS(j)$	Number of discrete dimensions $s(j)$ in the process of stage j.
NT	Number of time periods t.

Parameters

λdvt	Negative deviation on NPV penalization parameter.
λqns	Non-satisfied demand penalization parameter.
λslk	Capacity slack penalization parameter.
$prob_r$	Probability of scenario r.

Variables

ξ_r	NPV in scenario r.
$dvtn_r$	Negative deviation on the NPV in scenario r.
Qns_{itr}	Non-satisfied demand quantities of product i, (*robust*) in period t and scenario r.
$Cslk_{ijtr}$	Capacity slacks in each stage j, concerning totality of the *batches* of each product i, (*robust*) in period t and scenario r.

In addition, particular attention is paid to uncertainty modeling. Namely, product demands, Q_{itr} are uncertain and scenario dependent, this attribute is reflected in the full satisfaction of demand restrictions:

- In the deterministic model, the binding attribute for demand satisfaction directly implies the equipment sizing; however, while such an attribute can lead to larger sizes in the stochastic model too, the associated scenarios are typically presenting low probabilities.

- The produced quantities, W_{itr} are defined through a soft constraint, which simultaneously occurs with the definition of unmet or non-satisfied demand, Qns_{itr}; these recourse variables correspond to slack constraints, and such definition makes them available for penalization in the robust objective function.

$$W_{itr} \le Q_{itr} \text{, then}$$

$$W_{itr} + Qns_{itr} = Q_{itr} \text{ , } \forall i,t,r \tag{10.5}$$

Taking a softening procedure for demand satisfaction into account, similar procedures are developed in order to better formulate both the negative deviations to the NPV objective, $dvtn_r$, and the capacity slacks, $Cslk_{ijtr}$. For the former, by promoting linear relations and avoiding quadratic terms for solutions variability; for the latter, by defining these recourse variables through slack variables, in a similar procedure as described for unmet demands.

Thereafter, the Robust-DSBP problem is targeting NPV maximization within the flowshop configuration, while considering multiple machines in parallel at each stage and SPC and ZW policies. Beyond the expected value for NPV, the robust objective function incorporates penalty terms on solutions variability, non-satisfied demands, and capacities slack. These four items are represented, respectively, by the four sums that are formulated in the Robust-DSBP objective function:

$$\max \Phi = \sum_{r=1}^{NR} prob_r \xi_r - \lambda dvt \sum_{r=1}^{NR} prob_r.dvtn_r - \lambda qns \sum_{r=1}^{NR} \frac{prob_r}{NC.NT} \left(\sum_{i=1}^{NC} \sum_{t=1}^{NT} Qns_{itr} \right)$$

$$- \lambda slk \sum_{r=1}^{NR} \frac{prob_r}{M.NC.NT} \left(\sum_{j=1}^{M} \sum_{i=1}^{NC} \sum_{t=1}^{NT} Cslk_{ijtr} \right) \tag{10.6}$$

Note the NPV at each scenario, r, considers the expected return minus the investment costs, these two terms are represented by the respective sums in the relations shown in Equation 10.7:

$$\xi_r = \sum_{i=1}^{NC} \sum_{t=1}^{NT} ret_{itr} W_{itr} - \sum_{j=1}^{M} \sum_{s=1}^{NS(j)} \sum_{p=1}^{NP(j)} c_{jsp} y_{jsp} \text{ , } \forall r \tag{10.7}$$

10.4.1 Analytical Insights

Let the Robust-DSBP objective function be restricted to a special case with one single time period ($NT = 1$), and feature the following parameters and data:

- Penalization parameters are set to zero, $\lambda dvt = \lambda slk = 0$ (non-robust optimization); the related sums in the objective function (Equation 10.6) can thus be ignored, and

$$
\max \Phi = \sum_{r=1}^{NR} prob_r \xi_r - \lambda qns \sum_{r=1}^{NR} \frac{prob_r}{NC.NT} \left(\sum_{i=1}^{NC} \sum_{t=1}^{NT} Qns_{itr} \right) \quad (10.8)
$$

- Similar to the deterministic problem, only one time period is assumed $NT = 1$; therefore, index t is redundant and can be removed from the instance too.

- By defining $\lambda qns = bigM$ (a very large, majoring value), the unmet demands converge to zero, $Qns = 0$; then, the objective function presented at the relation in Equation 10.8 will be reduced to the expected value problem, as in the prior relation 10.1

$$
\max \Phi = \sum_{r=1}^{NR} prob_r \xi_r = E(\xi) \quad (10.9)
$$

- Additionally, note that by assuming $Qns_{itr} = 0$, $\forall i,t,r$, then the relations in Equation 10.5 will ensure the full satisfaction of product demands through the produced quantities:

$$
W_{itr} = Q_{itr}, \quad \forall i,t,r \quad (10.10)
$$

To summarize, the objective function for the Robust-DSBP problem is restricted to a specific instance, and the following can be ignored and removed: the time period index, t, because one single time period is considered; the sum sets concerning both the recourse variables negative deviations on NPV, $dvtn_r$, and capacity slacks, $Cslk_{ijtr}$; and also for the non-satisfied demands, Qns_{itr}, which consequently will be null and useless. Through a reduction approach and developing similar procedures for all the restriction sets in the robust problem, the stochastic EVP is thus obtained.

In concordance with the analytical approach in progress, Figure 10.1 shows the convergence of robust and stochastic objective values when the penalty parameter, λqns, successively increases. In addition, the sensitivity of the objective function with the penalty parameter shows model robustness, with the expected values for EVP and robust cost, $NPVrob$, altering rapidly with the λqns evolution, and a flat level is reached in both cases.

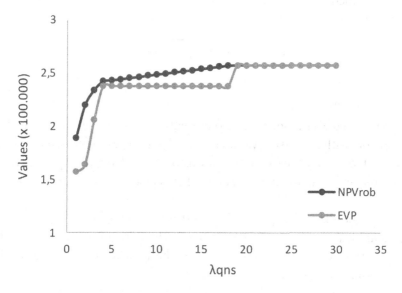

FIGURE 10.1   Model robustness: Variation of robust values (10^5 €) with penalty parameter, λqns

Addressing the stochastic EVP, diverse results of interest can be obtained, namely for validation purposes of numerical results.

- A specific instance with one single scenario is assumed; since $NR = prob_r = 1$, the optimal solution is reduced to the solution for the deterministic DSBP problem.

- When assuming a number of scenarios, with the same data and uniform distribution, the stochastic instance can be compared to the deterministic DSBP problem too; in fact, the instance corresponds to one single state of nature with the expected values for uncertain coefficients and parameters, which is replicated onto a number of scenarios with equal probabilities.

- In these two instances, the optimal value for NPV remains the same for all scenarios, and the negative deviation on NPV shall take zero value, $dvtn = 0$; then, the associated relations can be removed from the formulations at hand.

- For both of the described instances, the following relation is thus obtained:

$$\max \; \Phi \; = E(\xi) = \xi \sum_{r=1}^{NR} prob_r = \xi \tag{10.11}$$

10.4.2 Graphical Analysis

Graphical representations are also presented for the variation of different estimators with increasing values for penalization parameters, namely, for i) expected value of negative deviations on NPV, $Edvt$; ii) expected value of non-satisfied demand, $Ensd$; and iii) expected value of capacity slacks, $Eslk$.

First, the discussed probabilistic estimators will be given, and then the graphical analysis will be developed. Then, in concordance with the robust objective function in relation 10.6, the associated definitions for the estimators in the penalization terms follow.

- The expected value for negative deviations on NPV, $Edvt$, is utilized to penalize solutions variability:

$$Edvt = \sum_{r=1}^{NR} prob_r . dvtn_r \tag{10.12}$$

The negative deviation $dvtn$ is a linear measure that penalizes only the solutions with NPV below the NPV's expected value:

$$dvtn_r \geq \sum_{r'=1}^{NR} \left(prob_{r'} \xi_{r'} \right) - \xi_r \geq 0 , \; \forall r \tag{10.13}$$

- The expected value for non-satisfied demand, $Ensd$, is utilized to penalize the violations on soft restrictions for product demands:

$$Ensd = \sum_{r=1}^{NR} \frac{prob_r}{NC.NT} \left(\sum_{i=1}^{NC} \sum_{t=1}^{NT} Qns_{itr} \right) \tag{10.14}$$

The non-satisfied demands, Qns, are calculated through the difference between the demanded quantities, Q_{it}, and the produced quantities, W_{it}, and they correspond to the slack variables of the relations shown in Equation 10.5.

- The expected value for capacity slackness, $Eslk$, is utilized to penalize the over-sizing in terms of the equipment volumes, which are not fully utilized in the production batches:

$$Eslk = \sum_{r=1}^{NR} \frac{prob_r}{M.NC.NT} \left(\sum_{j=1}^{M} \sum_{i=1}^{NC} \sum_{t=1}^{NT} Cslk_{ijtr} \right) \tag{10.15}$$

The capacity slackness, $Cslk$, in each instance is also defined as slack variables that allow violations of soft restrictions associated with equipment sizes, in a similar way as developed for the non-satisfied demands, Qns.

In Figure 10.2, the sensitivity of the variability estimator, $Edvt$, with the evolution of penalty parameter for solutions deviation, λdvt, is provided. Then:

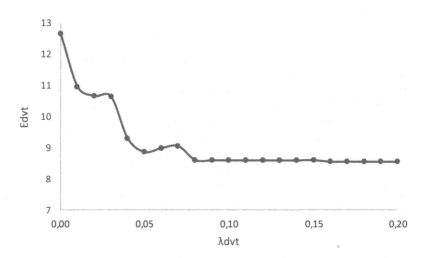

FIGURE 10.2   Solution robustness: Variation of expected value for negative deviation, $Edvt$, with penalty parameter, λdvt

- For λdvt penalty values in the 0–2 range, the estimator, $Edvt$, reduces rapidly with the penalty evolution and a flat level is reached.

- The estimator sensitivity is adequate for robustness promotion on solutions; that is, the rapid variation of $Edvt$, as well as the low plateau obtained, are both contributing for robustness on Robust-DSBP solutions.

- Assuming the λdvt penalty takes a value, for example 0.10, the expected value for solutions variability is decreasing quickly by about 30–40%, while the value reached for the estimator is supposed to represent a lower limit.

In Figure 10.3, the sensitivity of capacity slacks estimator, $Eslk$, with the evolution of penalty parameter for non-satisfied quantities, λqns, is shown. Then:

- For λqns penalty values in the 0–30 range, the estimator, $Eslk$, increases rapidly and then it keeps within a 10% range, approximately, $Eslk = 1500 \pm 150$ dm^3.

- The adequate sensitivity of the estimator, $Eslk$, that is, the rapid variation and the plateau obtained, correspond to robustness on the Robust-DSBP modeling.

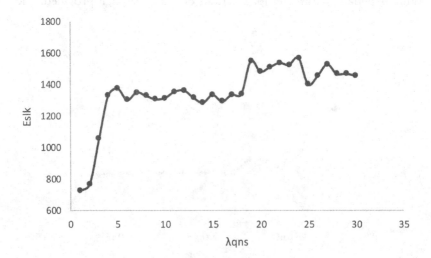

FIGURE 10.3   Model robustness: Variation of expected value for slack capacities, $Eslk$ (dm^3), with penalty parameter, λqns

- Assuming the λqns penalty takes a value, for example 5, the expected value for capacity slackness quickly increases to the plateau range, with the value reached representing an upper limit for the estimator under analysis.

Figure 10.4 displays the sensitivity of the non-satisfied demands estimator, *Ensd*, with the evolution of penalty parameter for non-satisfied quantities, λqns. Then:

- Again for λqns in the 0–30 range, the estimator, *Ensd*, decreases rapidly with the penalty evolution, a flat level is reached, and the estimator is nullified.

- The adequate sensitivity of estimator *Ensd*, that is, the rapid variation and the plateau obtained, also represents robustness on the Robust-DSBP modeling.

- For example, assuming again λqns takes a value of 5, the expected value for non-satisfied demand is quickly decreasing to reach a very low plateau, just before the estimator reaches its lower limit, zero.

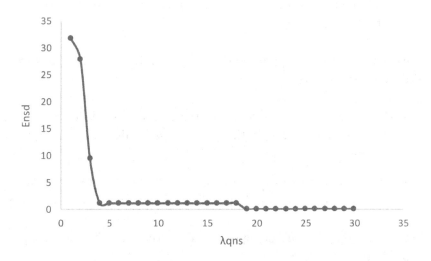

FIGURE 10.4   Model robustness: Variation of the expected value for non-satisfied demand, *Ensd* (10^3 kg), with penalty parameter, λqns

10.4.3 Notes on the Robust Approach

Nowadays, with the existing uncertainty and high levels of risk in several dimensions (*e.g.*, political, financial, economical, and environmental), investment and innovation in industry-based supply chains benefit from a sound understanding of DSBP problems, the related designing/planning/scheduling sub-problems, and the implementation of dedicated computer-aided tools.

Through a RO approach, the deterministic DSBP problem is generalized to a stochastic framework, integrating two phases for decision-making and robustness promotion: the Robust-DSBP problem. Within an uncertain environment, the RO approach allows significant investment cost decrease (from 8% to 20%) in the face of a slight relaxation on the satisfaction of uncertain demands (from 1% to 6%).

The generalization directed to the Robust-DSBP problem considers two important aspects:

- A stochastic framework, with the sizing of batch processes associated with the optimal satisfaction of uncertain demands; this static approach focuses on a single time period where investment costs are minimized.

- A dynamic approach, considering simultaneously demand uncertainty and multi-period time horizons, where the objective is NPV's robust maximization.

These characteristics support RO's usefulness. The characteristics were also shown to support:

- Model robustness, by checking the estimator's adequacy for the evolution of penalties, both for non-satisfied demand and capacity slackness.

- Solutions robustness, resulting in the stability configurations for the selected equipment with the penalty evolution for solution variability.

Recent developments are being observed for these DSBP problems, particularly, in two senses:

- Successively more realistic formulations are developed, integrating either additional degrees of freedom or uncertainty treatment; these enhancements are making the problems resolution more and more difficult.

- However, solution procedures are considered at the very beginning of the DSBP modeling phase; pertinent topics utilized in resolution procedures of other difficult problems, such as large networks or graphs, heuristics, or approximation schemes, are introduced too.

Adequate computer-aided tools will contribute to promote a lean status for industry-based supply chains, to better analyze data and foster results, as well as to enhance the design procedures. In this way, the generalized and Robust-DSBP problem simultaneously considered the long-term investment and short-term scheduling for flowshop batch processes.

It is also known that the robust problem Robust-DSBP is NP-hard in the strong sense. Thus polynomial algorithms can be ignored, while approximation procedures can be good options.

- These analytical studies also support heuristics and numerical implementation of DSBP problems.

- The treatment of reasonable dimension instances for the Robust-DSBP problem is feasible, due to progress in hardware, solution methods, and implementation procedures.

- However, it is pertinent to focus on optimality issues, while promoting robustness in the face of medium and large horizon time uncertainties.

In the DSBP problem context, different approaches often appear in response to either external (*e.g.*, prices and demand for products) or internal variations (*e.g.*, technical parameters, processing times), as well as diverse methodologies and resolution methods.

- Through multi-parametric optimization, investment costs are minimized, including the number of equipment in discrete dimensions, the operations flexibility index, as well as the environmental impact.

- Due to the inherent hardness of combinatorial optimization methods, carrying out a Pareto evaluation for the solutions under analysis is a suitable, and frequent, approach too.

- In the initial phase of DSBP developments, the importance of simultaneously optimizing both the scheduling of batch processes and environmental impacts is well recognized.

10.5 CONCLUDING REMARKS

The treatment of uncertainty is enhanced through RO approaches, which allow decision-makers to develop multiple scenarios and evaluate them in a probabilistic way. Multiple states of nature, in addition to related probabilities, are considered, and adequate stochastic treatments are thus developed.

Stochastic treatments surpass the typical expectation value; in fact, the generalization of deterministic models towards robustness improves decision-making and the related attributes, both qualitatively and quantitatively. The two phases framework for decision-making, or even considering multiple phases, is formally expressed through 2SSP; then, the recourse phase allows decision-makers to mitigate deviations in ways suitable for coping with uncertainty.

In addition, RO is able to promote the sensitivity of robust objective values with the alteration of probabilistic estimators through adequate penalization parameters; that is, with a fine-tuned alteration, the penalty parameters can drive a significant improvement in the robust estimators. Then, robust models and solutions can both be obtained through RO approaches.

Summary tables showing the main approaches and methods in this book were presented too. Overall:

- First, **introductory tools** and first notes on decision processes were presented.

- Second, **LA** provided simple solution methods, but simplicity comes at the cost of major shortcomings for decision support.

- **LP** broadened the results space, translating into qualitative and quantitative enhancements, by adding degrees of freedom and treating sensitive factors.

- **Duality** complemented and expanded on LP; a complementary and second decision-maker is assumed, while insights into resources utilization were provided.

- The generalization approach followed with **calculus optimization**, in this way enhancing the knowledge for decision-makers and connecting LP dual values with Lagrange multipliers.

- **LP optimality analysis** provided new information sources that enhance responses to uncertainty by systematically addressing coefficients and RHS parameter sensitivity, parametrically mapping solution alterations, or extending the alternatives space with linear combinations of optimal solutions.

- **ILP** solutions were directly translated to the real world, providing better modeling tools for decision-makers; however, a reduction approach was followed, and those improvements were accompanied by quantitative bounding.

- **Game Theory** deepened the decision framework; in this book, a second decision-maker is assumed and both mixed strategies and dominant strategies were treated, while the complementary LP formulations were developed.

- When facing **uncertainty** and trying to cope with the limitations of optimization methods, decision-maker behavior was extended by providing multiple perspectives for decision-makers beyond the *pessimistic* point of view.

- **RO** extended the decision process into diverse phases, two or more, and allowed proactive actions on data uncertainties; for that, generalizing deterministic optimization problems is a suitable approach to promote robustness both on solutions and models.

A case study was described to address a DSBP problem and a corresponding robust generalization was outlined. DSBP is a well-studied problem, due to its wide application in multiple industry sectors, for example, in agri-food, fine chemicals, food specialties, and pharmaceutical. Multi-criteria decision aiding/making (MCDA/M) tools were also utilized to address industry-based supply chains, dealing with uncertainty, coping with computational issues, and developing RO models.

The treatment of uncertainty considers problem specificities, promotes robustness both for the solution and model, and enhances the evaluation of technical–economic parameters. The analytical studies contributed to more detailed problems and models, balancing their benefits and limitations, directing the robust generalization, and promoting the validation of computational results of partial and final solutions.

Using RO approaches, topics located at the border of applied sciences and industrial production can be integrated through specific indexes related to social impacts, managerial economics, and environmental issues. In addition, multiple similarities can be found in real world applications, and optimization and decision-making improvements are also under consideration in many international initiatives and programs.

Selected References

Barbosa-Póvoa, A.P., Corominas, A., and Miranda, J.L. (Eds.), *Optimization and Decision Support Systems for Supply Chains.* Springer, Cham (2016). ISBN: 978-3-319-42421-7

Barbosa-Póvoa, A.P., and Miranda, J.L. (Eds.), *Operations Research and Big Data, Proceedings of IO2015-XVII Portuguese OR Conference.* Springer, Cham (2015). ISBN: 978-3-319-24152-4

Belien, J., Ittmann, H.W., Laumanns, M., Miranda, J.L., Pato, M.V., and Teixeira, A.P., (Eds.), *Advances in Operations Research Education - European Studies,* Springer International Series, Cham (2018). ISBN: 978-3-319-74103-1

Ben Amor, S., Teixeira, A., Miranda, J.L., and Aktas, E. (Eds.), *Advanced Studies in MCDM.* CRC Press, Boca Raton, FL (December-2019). ISBN: 978-1-138-74388-5

Biegler, L.T., Grossmann, I.E., and Westerberg, A.W., *Systematic Methods of Chemical Process Design.* Prentice-Hall, Hoboken, NJ (1997). ISBN: 978-0134924229

Eppen, G.D., Gould, F.J., and Schmidt, C.P., *Introductory Management Science.* Prentice Hall, Hoboken, NJ (1998). ISBN: 0-13-503582-1

Garey, M.R., and Johnson, D.S., *Computers and Intractability: A Guide to the Theory of NP-Completeness.* W. H. Freeman & Co., New York, NY (1990). ISBN: 978-0-7167-1045-5

Hillier, F.S., and Liebermann, G.J., *Introduction to Operations Research.* McGraw Hill, New York, NY (2009). ISBN: 978-0-07-132483-0

Liebowitz, J., *Developing Informed Intuition for Decision-Making.* CRC Press, Boca Raton, FL (2019). ISBN: 978-0-429-29809-7

Murty, K.G. (Ed.), *Case Studies in Operations Research,* Springer International Series, Springer, New York, NY (2015). ISBN: 978-1-493-91007-6

Ravindran, A.R. (Ed.), *Operations Research and Management Science Handbook.* CRC Press, Boca Raton, FL (2007). ISBN: 978-0-849-39721-9

Tavares, L.V., Oliveira, R.C., Themido, I.H., and Correia, F.N., *Investigação Operacional.* McGraw Hill de Portugal, Lisbon (2006). ISBN: 972-8298-08-0

Wainwright, K., and Chiang, A., *Fundamental Methods of Mathematical Economics.* McGraw Hill, New York, NY (January 2005). ISBN: 978-00 70109100

Index

Note: Page numbers in italic and bold refer to figures and tables, respectively.

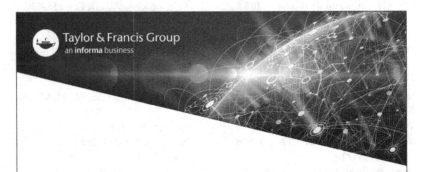

Printed in the United States
by Baker & Taylor Publisher Services